MATHEMATICS DEPARTMENT
EDMONTON COUNTY SCHOOL

3
Modern Mathematics for Schools

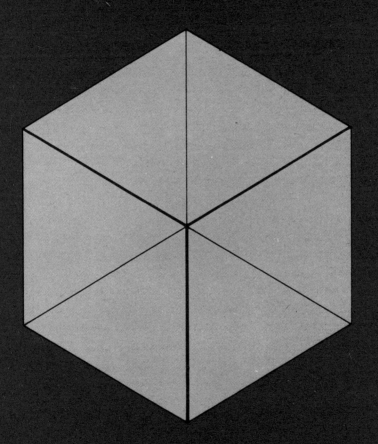

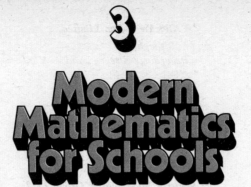

Second Edition
Scottish Mathematics Group

Blackie

Chambers

Blackie & Son Limited
Bishopbriggs · Glasgow
5 Fitzhardinge Street · London W1

W & R Chambers Limited
11 Thistle Street · Edinburgh 2
6 Dean Street · London W1

© *Scottish Mathematics Group 1972*
First Published 1972

All Rights Reserved.
No part of this publication may be reproduced,
stored in a retrieval system, or transmitted,
in any form or by any means,
electronic, mechanical, recording or otherwise,
without prior permission of the Publishers

Designed by James W. Murray

International Standard Book Numbers
Pupils' Book
Blackie 0 216 89408 5
Chambers 0 550 75913 1
Teachers' Book
Blackie 0 216 89409 3
Chambers 0 550 75923 9

Printed in Great Britain by
Butler & Tanner Limited · Frome and London
Set in 10pt Monotype Times Roman

Members associated with this book

W. T. Blackburn
Dundee College of Education

Brenda I. Briggs
Formerly of Mary Erskine School for Girls

W. Brodie
Trinity Academy

C. Clark
Formerly of Lenzie Academy

D. Donald
Formerly of Robert Gordon's College

R. A. Finlayson
Allan Glen's School

Elizabeth K. Henderson
Westbourne School for Girls

J. L. Hodge
Madras College

J. Hunter
University of Glasgow

T. K. McIntyre
High School of Stirling

R. McKendrick
Langside College

W. More
Formerly of High School of Dundee

Helen C. Murdoch
Hutchesons' Girls' Grammar School

A. G. Robertson
John Neilson High School

A. G. Sillitto
Formerly of Jordanhill College of Education

A. A. Sturrock
Grove Academy

Rev. J. Taylor
St. Aloysius' College

E. B. C. Thornton
Bishop Otter College

J. A. Walker
Dollar Academy

P. Whyte
Hutchesons' Boys' Grammar School

H. S. Wylie
Govan High School

Only with the arithmetic section

R. D. Walton
Dumfries Academy

Book 1 of the original series *Modern Mathematics for Schools* was first published in July 1965. This revised series has been produced in order to take advantage of the experience gained in the classroom with the original textbooks and to reflect the changing mathematical needs in recent years, particularly as a result of the general move towards some form of comprehensive education.

Throughout the whole series, the text and exercises have been cut or augmented wherever this was considered to be necessary, and nearly every chapter has been completely rewritten. In order to cater more adequately for the wider range of pupils now taking certificate-oriented courses, the pace has been slowed down in the earlier books in particular, and parallel sets of A and B exercises have been widely introduced. The A sets are easier than the B sets, and provide straightforward but comprehensive practice; the B sets have been designed for the more able pupils, and may be taken in addition to, or instead of, the A sets. Occasionally a basic exercise, which should be taken by all pupils, is followed by a harder one on the same work; in such a case the numbering is, for example, Exercise 2 followed by Exercise 2B. It is hoped that this arrangement, along with the 'Graph Workbook for Modern Mathematics', will allow considerable flexibility of use, so that while all the pupils in a class may be studying the same topic, each pupil may be working examples which are appropriate to his or her aptitude and ability.

Each chapter is backed up by a summary, and by A and B revision exercises; in addition, cumulative summaries and exercises have been introduced at the end of alternate books. A new feature is the series of Computer Topics from Book 4 onwards. These form an elementary introduction to computer studies, and are primarily intended to give pupils some appreciation of the applications and influence of computers in modern society.

Books 1 to 7 provide a suitable course for many modern Ordinary Level and Ordinary Grade syllabuses in mathematics, including the University of London GCE Syllabus C, the Associated Examining Board Syllabus C, the Cambridge Local Syndicate Syllabus C, and the Scottish Certificate of Education. Books 8 and 9 complete the work for the Scottish Higher Grade Syllabus, and provide a good preparation for all Advanced Level and Sixth Year Syllabuses, both new and traditional.

Related to this revised series of textbooks are the *Modern Mathematics Newsletters*, the *Teacher's Editions* of the textbooks, the *Graph Workbook for Modern Mathematics*, the *Three-Figure Tables for Modern Mathematics*, and the booklets of *Progress Papers for Modern Mathematics*. These new Progress Papers consist of short, quickly marked objective tests closely connected with the textbooks. There is one booklet for each textbook, containing A and B tests on each chapter, so that teachers can readily assess their pupils' attainments, and pupils can be encouraged in their progress through the course.

The separate headings of Algebra, Geometry, Arithmetic, and later Trigonometry and Calculus, have been retained in order to allow teachers to develop the course in the way they consider best. Throughout, however, ideas, material and method are integrated *within* each branch of mathematics and *across* the branches; the opportunity to do this is indeed one of the more obvious reasons for teaching this kind of mathematics in the schools—for it is *mathematics* as a whole that is presented.

Pupils are encouraged to find out facts and discover results for themselves, to observe and study the themes and patterns that pervade mathematics today. As a course based on this series of books progresses, a certain amount of equipment will be helpful, particularly in the development of geometry. The use of calculating machines, slide rules, and computers is advocated where appropriate, but these instruments are not an essential feature of the work.

While fundamental principles are emphasized, and reasonable attention is paid to the matter of structure, the width of the course should be sufficient to provide a useful experience of mathematics for those pupils who do not pursue the study of the subject beyond school level. An effort has been made throughout to arouse the interest of all pupils and at the same time to keep in mind the needs of the future mathematician.

The introduction of mathematics in the Primary School and recent changes in courses at Colleges and Universities have been taken into account. In addition, the aims, methods, and writing of these books have been influenced by national and international discussions about the purpose and content of courses in mathematics, held under the auspices of the Organization for Economic Co-operation and Development and other organizations.

The authors wish to express their gratitude to the many teachers who have offered suggestions and criticisms concerning the original series of textbooks; they are confident that as a result of these contacts the new series will be more useful than it would otherwise have been.

Algebra

1 Relations, Mappings and Graphs 3

Relations between two sets; a relation as a set of ordered pairs; mappings; one-to-one correspondence; graphs.

2 Operations on Integers and Rational Numbers 18

Addition of integers; subtraction of integers; using the addition and subtraction of integers; multiplication of integers; using the multiplication of integers; the associative law of multiplication for integers; the distributive law for integers; the set of rational numbers—addition and subtraction; multiplication and division.

3 Equations and Inequations in One Variable 40

Equations and inequations; adding the same number to each side of an equation; multiplying each side of an equation by the same number; using negatives and reciprocals to solve equations; adding the same number to each side of an inequation; the use of set-builder notation; multiplying each side of an inequation by the same number; using negatives and reciprocals to solve inequations; equations and inequations with fractions; applications to problems.

Revision Exercises	60

Geometry

1 Reflection 73

Symmetry; symmetry about an axis; reflection in a line;
coordinates; the rhombus; some constructions;
the kite; some problems.

2 The Parallelogram 96

Half turns and parallel lines; the parallelogram;
parallelogram tiling;
angles associated with parallel lines—corresponding angles and
alternate angles; the areas of a parallelogram.

3 Locus, and Equations of a Straight Line 114

The idea of a locus; sets of points;
intersection of sets; paths of moving objects;
equations of a straight line.

Revision Exercises 133

Arithmetic

1 Social Arithmetic 1 — 147
Money in the home—electricity and gas accounts,
ready reckoners, discount;
money in the bank—current accounts, deposit accounts,
interest, finding the rate of interest;
money in business—profit and loss;
more percentages—common fractions, decimal fractions and
percentages; using percentages.

2 Ratio and Proportion — 165
Ratio; direct proportion—unitary method and ratio method of
calculation; maps and plans;
inverse proportion—product method and ratio method of
calculation; graphs; miscellaneous questions.

3 Introduction to Probability — 185
Experiment; theory; calculating expected frequencies;
probability of certain success and certain failure;
combining outcomes—mutually exclusive and independent outcomes.

4 Time, Distance, Speed — 203
Timetables; distance–time graphs;
time–distance–speed calculations.

Revision Exercises — 214

Answers — 227

Notation

Sets of numbers

Different countries and different authors give different notations and definitions for the various sets of numbers. In this series the following are used:

E The universal set

ϕ The empty set

N The set of natural numbers $\{1, 2, 3, ...\}$

W The set of whole numbers $\{0, 1, 2, 3, ...\}$

Z The set of integers $\{..., -2, -1, 0, 1, 2, ...\}$

Q The set of rational numbers

R The set of real numbers

The set of prime numbers $\{2, 3, 5, 7, 11, ...\}$

Algebra

Algebra

Note to the Teacher on Chapter 1

The concepts of relation and function (or mapping) are fundamental in mathematical analysis. In this chapter, the ideas of sets developed in Book 1 will be used to clarify and illuminate the usage of relations, not only in mathematics, but in everyday life. Intuitively, a relation defines some sort of association between elements of two sets, which need not be distinct, such that a pairing of elements is induced *from* one set *to* the other. This intuitive concept will form the basis of the investigation into what a relation is, as distinct from what it does.

Since the chapter is necessarily introductory, mathematical terms and technicalities are deliberately reduced to a minimum so that the pupil can concentrate on the ideas themselves. Diagrammatic representations of relations and mappings are therefore emphasized, and their use is encouraged throughout the chapter.

Section 1 opens with an informal discussion of a simple relation 'owned' which could form the basis of an introductory class lesson. The arrow diagram shows clearly the pairing of elements from one set to another. So far, it would appear that a relation is something that 'holds' between certain objects and fails to hold between other objects. Exercise 1 is designed with this idea in mind.

In *Section* 2, attention is directed to the ordered pairs in relations. Suppose we have a relation R (e.g. *'owned'* from a set A to a set B). Let x be a variable on A and let y be a variable on B. Using these variables and the relation R, we construct the open sentence $x \, R \, y$. If $a \in A$ and $b \in B$, the statement $a \, R \, b$ may be true or false. For example, 'Mary *owned* a transistor set' is true, but 'Mary *owned* a record player' is false. The set of all ordered pairs (a, b) such that $a \, R \, b$ is true is the solution set of the open sentence $x \, R \, y$. The relation R is defined by this set of ordered pairs. There is now motivation for the introduction of the Cartesian graph, as an alternative to the arrow diagram, for showing relations. The superiority of the Cartesian graph over the arrow diagram becomes evident when the sets have a comparatively large number of elements. It may perhaps seem unnatural to describe a relation by an open sentence $x \, R \, y$, or by a set of ordered pairs. However, in mathematics this interpretation has many advantages and is quite a reasonable one as the following example may show. Suppose that a pupil is asked to explain to a friend the meaning of 'is taller than'. He would probably begin by citing examples such as 'Peter is taller than Bill', in the hope that by giving a sufficient number of examples, his friend would understand the meaning of 'is taller than'.

Section 3 introduces the concept of a mapping as a special kind of relation. At this stage, there is a temptation to introduce technical words such as 'onto', 'many-one', domain, range, and so on, to sharpen descriptions of mappings. No attempt is made to do so however. Attention is concentrated on the mapping itself. The essential points to emphasize are that if A and B are two (non-empty) sets and there is a relation from A to B such that

(1) each element of A appears in the relation once, and
(2) each element of A is joined by one, and only one, arrow to some element of B,

then the relation is a mapping from A to B.

The pupil should also understand that in a mapping from a set A to a set B, not all the elements of B need appear in the mapping, and that several arrows may converge on any one element of B. Exercise 3 contains many questions which should help to reinforce the understanding of the basic properties of a mapping. A more sophisticated treatment of mappings will be given in Book 5.

Section 4 introduces a special kind of mapping, namely a one-to-one correspondence. Further instances of one-to-one correspondences will be met in the course, e.g. in proportion, in transformations in geometry, and in the theory of logarithms. It is useful to note that if two finite sets A and B are in one-to-one correspondence, A and B have the same number of elements. In other words, A and B have the same cardinal number.

The chapter concludes with an introduction to graphs of mappings which is complementary to the Geometry Chapter 3 on 'Locus'. Although the pupil is familiar only with the sets N, W, Z and Q, there are advantages in making an assumed extension of the rational numbers to the set R of real numbers. This can be done through plausible geometrical considerations. The problem of developing the set of real numbers as an extension of the set of rational numbers is rather difficult at this level of work. In Book 4, it will be demonstrated that irrational numbers can be represented by points on the number line, which suggests that there is a one-to-one correspondence between *all* points on the number line and the set of real numbers.

(facing page 3)

1 Relations, Mappings and Graphs

1 Relations between two sets

Four girls, Mary, Jane, Betty and Nan, were asked if they owned a transistor set, record player or tape recorder. Their answers showed that:

a Mary and Jane each owned a transistor set (TS).
b Betty and Nan each owned a record player (RP).
c Mary and Betty each owned a tape recorder (TR).

From this information, can you answer the following questions?

1 Which girls owned all three instruments?
2 Which girls owned two instruments?
3 Which girls owned only one instrument?

Notice that there are two sets, a set of girls {Mary, Jane, Betty, Nan} which we will denote by A, and a set of instruments {TS, RP, TR} which we will denote by B.

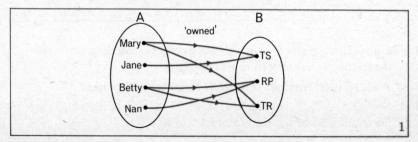

Algebra

Figure 1 shows a way of analysing the given information, and answering the questions asked. Each time a girl *owned* a particular instrument we draw an arrow *from* the girl *to* the instrument. Figure 1 is called *an arrow diagram*.

By following the arrows we see that:

1 None of the girls owned all three instruments.
2 Mary and Betty each owned two (which two?).
3 Jane and Nan each owned one (which ones?).

Figure 1 shows a relation *from* set A *to* set B, the relation being 'owned'. Notice that a relation has *sense*, i.e. a direction in which it goes, as indicated by the arrows in the diagram.

A *relation* from set A to a set B is a pairing of elements of A with elements of B.

Exercise 1

1 Relations can be described in words. For example: 'is the brother of', 'is half of', 'lives in the same street as'. Write down four more examples.

2 Two sets of numbers, P and Q, are listed in Figure 2. Copy and complete the arrow diagram to show the relation *is greater than* from set P to set Q.

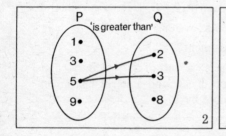
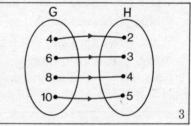

3 Using the same sets as in question *2*, draw an arrow diagram to show the relation *is less than* from set P to set Q.

4 What relation from set G to set H is illustrated in Figure 3?

5 $S = \{0, 1, 2, 5\}$, $T = \{1, 2, 3, 4, 6\}$. Make an arrow diagram to show the relation *is one less than* from set S to set T.

Relations between two sets

6 Construct an arrow diagram to show the relation *is a factor of* from set $A = \{2, 3, 5, 7, 11\}$ to set $B = \{1, 6, 12, 17, 30, 35\}$.

7 Mr Gray is the father of Gordon. Mr Ford is the father of Neil, James and Pauline. Mr Monteith is the father of Margaret.

Copy Figure 4, and complete the arrow diagram to show the relation *is the father of* from the set of fathers to the set of children.

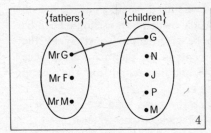
4

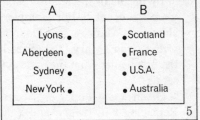
5

8 Using the same sets as in question 7, draw an arrow diagram to show the relation *is the son or daughter of* from the set of children to the set of fathers.

9 a Copy Figure 5 and draw an arrow from each town in set A to the country in set B in which it is situated.

  b Complete: 'Figure 5 shows the relation ... . from set A to set B.'

10 Mary and Jane are both clever. Jean and Jane are both tall. Mary and Jane are both fair.

  a List the set G of girls and the set C of their characteristics.
  b Draw an arrow diagram relating each girl to her characteristics.
  c Which girl is both tall and clever?

11 Bob and Ian are tall. Ian and Fred are dark. Bob and Jim are handsome. List the sets of boys and their characteristics, and draw an arrow diagram relating each boy to his characteristics.

  a Who is both tall and handsome?
  b Who is neither tall nor handsome?
  c Who is tall but not handsome?
  d Who is tall and dark?
  e Are any of the boys tall, dark and handsome?

Algebra

2 A relation as a set of ordered pairs

In Section 1 we had a set of girls $A = \{$Mary, Jane, Betty, Nan$\}$, and a set of instruments $B = \{$TS, RP, TR$\}$.

By means of the relation *owned* we linked members of set A with members of set B. For example, Mary owned a transistor set can be shown as Mary $\rightarrow$ TS. This can be given more concisely by the *ordered pair* (Mary, TS). The elements in the pair are *ordered* since the first comes from the first set (A), and the second from the second set (B).

Given $x \in A$ and $y \in B$, consider the open sentence x *owned* y.
Replacing x by 'Mary' and y by 'TS' gives a true sentence.
Replacing x by 'Mary' and y by 'RP' gives a false sentence.

The set of all ordered pairs (x, y) which give true sentences, i.e. the solution set of the open sentence, is:

$\{$(Mary, TS), (Mary, TR), (Jane, TS), (Betty, RP), (Betty, TR),
(Nan, RP)$\}$

This set of ordered pairs defines the relation *owned* from set A to set B.

Figure 6 shows a Cartesian graph of the relation; the coordinates of the points are the elements of the ordered pairs in the relation.

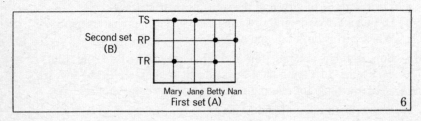

Exercise 2

1   Figure 7 shows the relation *plays* for the two sets indicated.
   a   Express the relation as a set of ordered pairs. (Use suitable initial letters.)
   b   Draw a Cartesian graph of the relation.

A relation as a set of ordered pairs

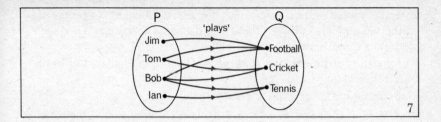

2    $A = \{2, 3, 5, 6\}$ and $B = \{1, 2, 3, 4, 5, 6\}$.
- **a** Show the relation *is a factor of* from A to B by an arrow diagram.
- **b** Write down the relation as a set of ordered pairs.

3   A relation between two sets is given by the set of ordered pairs: $\{(-1, 2), (1, 4), (3, 6), (5, 8), (7, 10)\}$.
      List the elements of the two sets, and describe in words a possible relation from the first set to the second set.

4    $X = \{0, 2, 4, 6, 8\}$ and $Y = \{0, 1, 2, 3, 4, 5\}$.
- **a** If $x \in X$ and $y \in Y$, list the set of ordered pairs in the relation *x is double y*.
- **b** Illustrate the relation by means of a Cartesian graph.

5    $S = \{a, e, i, o, u\}$ and $T = \{f, n, s, t\}$.
      If $x \in S$ and $y \in T$ list the set of ordered pairs in the relation *x followed by y spells a two-letter English word*, e.g. (a, n) is in the relation.

6    $A = B = \{1, 2, 3, 6\}$.
- **a** Copy and complete Figure 8 for the relation *is a factor of* from set A to set B.
- **b** Express the relation as a set of ordered pairs.
      Note. When two sets A and B are equal we refer to a relation *on A*.

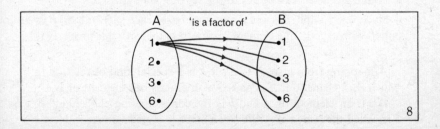

Algebra

7 *a* List the set of ordered pairs which describe the relation *is less than* on the set $A = \{1, 2, 3, 4, 5\}$ (i.e. from the set A to the set A). Hence draw a Cartesian graph of the relation.

   b Repeat *a* for the relation *is equal to*. Use the same diagram, but a different colour for the graph.

   c Repeat *a* for the relation *is greater than*.

8 $A = \{-1, 0, 1\}$ and $B = \{1, 2, 3, 4\}$. It is required to find the set of ordered pairs such that every element of A is paired with every element of B.

   a Copy and complete this table to find all such pairs.

   b Draw a Cartesian graph of the relation between the two sets.

	1	2	3	4
−1	(−1, 1)	(−1, 2)	.	.
0	.	.	.	(0, 4)
1	.	.	(1, 3)	.

9 Four teams p, q, r, s in a local football league have to play each other both at home and away. Using a table as in question *8* if you wish, list the set of games that have to be played, in the form of ordered pairs such as (p, q).

10 A relation R is defined by $\{(\tfrac{1}{2}, 8), (1, 4), (2, 2), (4, 1), (8, \tfrac{1}{2})\}$.

   a List the set of first members of the pairs, and the set of second members of the pairs. Describe in words a possible relation from the first set to the second.

   b Draw a Cartesian graph of the relation R, and draw a smooth curve through the points.

3 Mappings

The arrow diagram in Figure 9 illustrates the relation *takes shoe size* from the given set P of pupils to the set S of shoe sizes.

Since each pupil has only one shoe size, each element of P is paired with exactly one element of S. A relation with this property is called a *mapping*.

A mapping from a set A to a set B is a special kind of relation in which *each* element of A is paired with exactly one element of B.

If a is an element of A and b is the corresponding element of B, b is called the *image* of a. We say *a maps to b*, and write $a \to b$.

Mappings

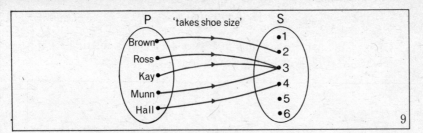

For example, from Figure 9, Brown → 2, and 2 is the image of Brown.

You have already met this idea in the geometry course. It also occurs in geographical map-making, where the towns are mapped to dots on paper.

Note that in the set of ordered pairs of a mapping, each element of A appears as the first member in only one pair; (2, 4) and (2, 7) could not appear as ordered pairs in a mapping.

To sum up, for a *mapping* from A to B we require:

a two sets A and B

b a relation which links *each* element of A to *exactly one* element of B

Example. Which of the relations shown in Figure 10 are mappings?

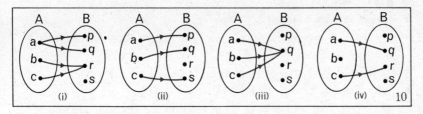

(i) is not a mapping since a is linked to two elements in B.
(ii) and (iii) are both mappings.
(iv) is not a mapping since b is not linked to an element in B.

Algebra

Exercise 3

1. Each of the arrow diagrams in Figure 11 shows a relation from a set P to a set Q. Which of these relations are mappings?

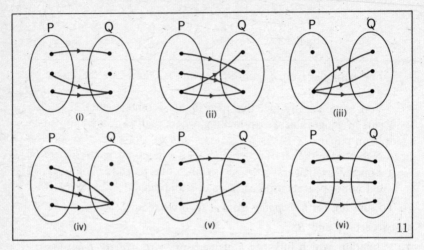

2. $X = \{a, b, c\}$ and $Y = \{1, 2, 3, 4\}$.
   a. Make an arrow diagram to show the mapping m defined by $a \to 1$, $b \to 2$, $c \to 2$.
   b. List m as a set of ordered pairs.

3. X and Y are the same sets as in question 2. A relation R from X to Y is such that $a \to 3$, $b \to 2$, $b \to 4$, $c \to 1$.
   a. Show R as a set of ordered pairs and as an arrow diagram.
   b. Why is R not a mapping from X to Y?

4. $S = \{1, 2, 3, 4\}$ and $T = \{8, 9\}$. A relation R takes each odd number in S to 8, and each even number in S to 9.
   a. List R as a set of ordered pairs, and draw an arrow diagram.
   b. Is R a mapping from S to T?

5. x and y are variables on the set $S = \{1, 2, 3, ..., 20\}$. A relation on S is defined by the open sentence: x is three times y.
   a. List the set of ordered pairs (x, y) which give the relation.
   b. Is this relation a mapping? Give a reason for your answer.

6. $K = \{1, 2, 3, 4, 5, 6\}$, and $n \in K$. A mapping f is such that if n is odd, $n \to 1$, and if n is even, $n \to \tfrac{1}{2}n$.

One-to-one correspondence

    a Write down the images of 1, 2, 3, 4, 5, 6 under *f*.
    b Express *f* as a set of ordered pairs.

7 $E = \{1, 2, 3, ..., 12\}$ and $A = \{1, 3, 5, 7, 9\}$. A mapping *f* from *A* to *E* is such that $n \rightarrow n+2$. Replacing *n* by members of *A*, find the set of ordered pairs in *f*, e.g. $(1, 3) \in f$.

8 *a* Copy Figure 12, and complete the members of the second set (*the image set*) for the mapping *has as capital*.
   b Suggest an easy way of adapting the diagram to show the mapping *is capital of* from the set of capitals to the set of countries.

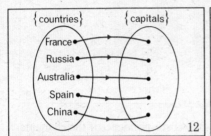 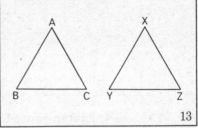

9   Figure 13 shows two equilateral triangles. List the six possible mappings of the vertices A, B, C to the vertices X, Y, Z, in which all three of the vertices X, Y, Z appear, e.g. A → X, B → Y, C → Z.

10 *a* Investigate by listing ordered pairs, or by making mapping diagrams, the number of mappings possible from set $A = \{a, b, c\}$ to set $B = \{p, q\}$.
   b Investigate the number of mappings possible from *B* to *A*.

---

4 One-to-one correspondence

Illustrations

   a Suppose each pupil in the class has one copy of *Modern Mathematics for Schools* Book 3. We say that there is a *one-to-one correspondence* between the set of pupils in the class and their set of mathematics books.
   b When numbered tickets are sold for a school dance, there is one

Algebra

ticket for each person attending. Here there is a *one-to-one correspondence* between the pupils attending and the tickets they hold.

c When the national flags are flown at the Olympic Games there is one particular flag for each competing nation. There is a *one-to-one correspondence* between the set of nations and the set of national flags.

In each case there is a two-way mapping which pairs all the elements in the sets. This is shown by double-headed arrows as follows:

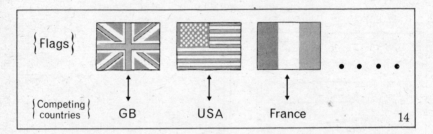

Two sets A and B are in one-to-one correspondence if the elements of A and B can be paired so that each element of A corresponds to one element of B, and each element of B corresponds to one element of A. Obviously the sets must contain the same number of elements.

Example. Show a one-to-one correspondence between the set $A = $ {vowels in the English alphabet} and the set $B = $ {odd numbers less than 10}.

$$A: a \quad e \quad i \quad o \quad u$$
$$\updownarrow \quad \updownarrow \quad \updownarrow \quad \updownarrow \quad \updownarrow$$
$$B: 1 \quad 3 \quad 5 \quad 7 \quad 9$$

Exercise 4

1 Which of the arrow diagrams in Figure 15 could give a one-to-one correspondence between the sets A and B?

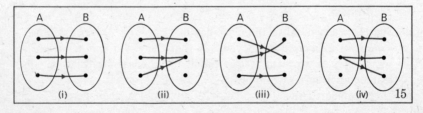

One-to-one correspondence

2 *a* Draw an arrow diagram to show a possible one-to-one correspondence between the sets $A = \{a, b\}$ and $P = \{p, q\}$.

b Draw another arrow diagram to show another possible one-to-one correspondence between the two sets.

3 One copy of a book costs 90p, two copies cost 180p, three copies cost 270p, and so on.

a What two sets are in one-to-one correspondence here?

b What is the cost of 7 books? How many books can be bought with £9?

4 An aircraft flies 500 km in 1 hour, 1000 km in 2 hours, 1500 km in 3 hours, and so on.

a What two sets are in one-to-one correspondence here?

b How far would the plane fly in 4 hours? How long would it take to fly 250 km?

5 Which of the following pairs of sets can be put in one-to-one correspondence?

a $A = \{0, 2, 4, 6\}$, $B = \{1, 3, 5, 7\}$.

b $P = \{$fingers on one hand$\}$, $Q = \{$days in a week$\}$.

c $X = \{a, b, c\}$, $Y = \{c, a, b\}$.

d $G = \{$natural numbers less than 4$\}$, $T = \{$the vertices of $\triangle ABC\}$.

6 Write down three examples from everyday life of one-to-one correspondence.

7 A shopkeeper wishes to mark the prices of his goods with letters, and chooses the code word *importance*. He sets up a one-to-one correspondence with the set of letters in this word and the set of numerals $\{0, 1, 2, 3, ..., 9\}$ thus:

$$\begin{array}{cccccccccc} I & M & P & O & R & T & A & N & C & E \\ \updownarrow & \updownarrow & \updownarrow & \updownarrow & \updownarrow & \updownarrow & \updownarrow & \updownarrow & \updownarrow & \updownarrow \\ 0 & 1 & 2 & 3 & 4 & 5 & 6 & 7 & 8 & 9 \end{array}$$

An article priced at £5·07 he marks T/IN.

a What would he put for (*1*) £1·32? (*2*) £0·76?

b What would these mean? (*1*) P/AI (*2*) M/MR.

8 In secret writing, VJG GPGOA KU KP UKIJV means THE ENEMY IS IN SIGHT. Set up a one-to-one correspondence for this code, and translate the coded message VJGA UGG OG.

9 In which of the following is there a one-to-one correspondence between the set of pupils in your class and the given set?

Algebra

    a {desks in your classroom}
    b {names on the class register}
    c {homes of pupils in your school}
    d {months of the year}
    e {dates of birth of pupils in the class}

10   Show by means of arrow diagrams that the sets $A = \{1, 2, 3\}$ and $B = \{p, q, r\}$ can be put in one-to-one correspondence in six ways.

11a Set up a one-to-one correspondence between the letters of the alphabet and the first twenty-six binary numbers, using five binary digits. The first three have been done for you:

        A 00001; B 00010; C 00011

    This correspondence expresses the letters of the alphabet in a five-binary-digit code, i.e. a 5 'bit' code, as we saw in Book 2.

  b Data can be fed into a computer on punched paper tape, a hole representing 1 and the absence of a hole representing zero. With a 5 'bit' code, we can show A as 00001, B as 00010, and so on. Use the correspondence to read the messages on the tapes shown in Figure 16.

5 Graphs

The graph of the mapping $x \to 2x$

Suppose that x is a variable on the set $A = \{0, 1, 2, 3, 4, 5\}$. Under the mapping $x \to 2x$ from the set A to the set W of whole numbers, $0 \to 0,\ 1 \to 2,\ 2 \to 4,\ 3 \to 6,\ 4 \to 8,\ 5 \to 10$.

    This gives the set of ordered pairs $\{(0, 0), (1, 2), (2, 4), (3, 6), (4, 8), (5, 10)\}$.

    Using these ordered pairs we can draw a Cartesian graph of the mapping $x \to 2x$, for $x \in \{0, 1, 2, 3, 4, 5\}$ as shown in Figure 17.

Graphs

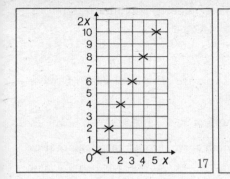
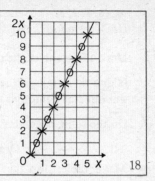

Now consider the graph of the mapping on the set of *all positive numbers and zero*.

We can show some of the ordered pairs in a table:

x	0	½	1	1½	2	2½	3	3½	4	4½	5
$2x$	0	1	2	3	4	5	6	7	8	9	10

The corresponding points on the graph are shown by crosses and circles in Figure 18.

In fact, points corresponding to all the ordered pairs of numbers given by the mapping $x \to 2x$ lie on the straight line which has been drawn in the figure.

Exercise 5

1. a Make a table for the mapping $x \to x+1$ from the set $\{0, 1, 2, 3, 4\}$ to the set of whole numbers.
   b Draw the graph of the set of ordered pairs of numbers.
   c Hence draw the graph of $x \to x+1$ on the set of all positive numbers and zero.

2. a Make tables for the mappings $x \to 2x$, $x \to 3x$ and $x \to 4x$ from the set $\{0, 1, 2, 3, 4\}$ to the set of whole numbers.
   b Draw the graphs of these mappings on the set of positive numbers and zero on the same sheet of squared paper.
   c Write the mapping alongside each graph, and write down three observations about the graphs.

3. Repeat question *2* for the mappings $x \to 2x$ and $x \to 2x+4$.

4. a Make a list of the ordered pairs for the mapping $x \to x^2$ from the

Algebra

set {0, 1, 2, 3, 4} to the set of whole numbers, and draw a graph.
b Draw a smooth curve through the points.

5 a Draw the graph of the mapping $x \to \dfrac{1}{x}$ for $x \in \{4, 3, 2, 1, \tfrac{1}{2}, \tfrac{1}{3}, \tfrac{1}{4}\}$, taking the origin in the centre of a page of squared paper. Draw a smooth curve through the points.

b On the same page, draw the graph of $x \to \dfrac{1}{x}$ for $x \in \{-4, -3, -2, -1, -\tfrac{1}{2}, -\tfrac{1}{3}, -\tfrac{1}{4}\}$. Draw a smooth curve through these points also.

6 a Make tables for the mappings $x \to x$, $x \to -x$, $x \to x+2$ and $x \to -x+2$ from the set $\{-5, -4, -3, -2, -1, 0, 1, 2, 3, 4\}$ to the set of integers.

b Draw graphs of these mappings on the set of positive numbers and zero on the same sheet of squared paper.

c Write the mapping alongside each graph, and write down what you notice about the graphs.

7    $A = \{(-4, -3), (-3, -2), (-2, -1), (-1, 0), (0, 1), (1, 2), (2, 3), (3, 4)\}$, and
$B = \{(-4, 5), (-3, 4), (-2, 3), (-1, 2), (0, 1), (1, 0), (2, -1), (3, -2)\}$.

a Construct graphs of the sets A and B of ordered pairs.
b Write down mappings which give A and B.

Summary

1. A *relation* from a set A to a set B is a pairing of elements of A with elements of B. If $B = A$, the relation is said to be *on* A.

2. A *relation* may be completely described by:
   (1) A *set of ordered pairs*.
   (2) The *solution set of an open sentence*, such as 'x owns y', which gives the ordered pairs in the relation.
   (3) An *arrow diagram* or a *Cartesian graph*.

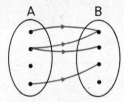 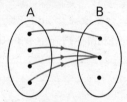

A relation which is not a mapping.   A relation which is a mapping.

3. A *mapping* from a set A to a set B is a relation in which every element of A is paired with exactly one element of B.

   The element of B is called the *image* of the element of A under the mapping.

4. Two sets A and B are said to be in *one-to-one correspondence* if the elements of A and B can be paired so that each element of A corresponds to exactly one element of B, and each element of B corresponds to exactly one element of A.

   Example. $A = \{1, 2, 3, ..., 12\}$, $B = \{\text{months of the year}\}$

1	2	3	4	5	
↕	↕	↕	↕	↕	...
Jan	Feb	Mar	Apr	May	

Operations on Integers and Rational Numbers

1 Addition of integers

In Book 2 we studied the set of integers
$$Z = \{..., -3, -2, -1, 0, 1, 2, 3, ...\}.$$
We saw how to add integers with the aid of a number line.
For example, Figure 1 illustrates $3 + (-9) = -6$

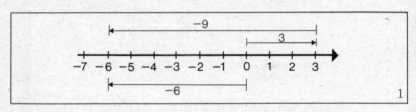

Exercise 1

Calculate the following, using a number line if you need to.

1	$3+(-2)$	2	$6+4$	3	$4+(-7)$	4	$-6+6$
5	$-4+3$	6	$-4+0$	7	$-5+5$	8	$-3+(-3)$
9	$-5+(-1)$	10	$0+(-2)$	11	$3+(-3)$	12	$-4+(-2)$
13	$-5+(-3)$	14	$-5+3$	15	$-8+10$	16	$11+(-9)$
17	$-11+9$	18	$-7+0$	19	$10+(-10)$	20	$12+(-7)$
21	$-4+(-4)$	22	$4+(-4)$	23	$-4+4$	24	$-9+(-9)$
25	$5a+(-2a)$	26	$7b+(-7b)$	27	$3c+(-6c)$	28	$c+(-5c)$
29	$-d+(-d)$			30	$-5p+(-7p)$		
31	$-6q+(-q)$			32	$-5r+2r$		

Note to the Teacher on Chapter 2

In Book 2, Algebra, Chapter 2, the set of integers was introduced as an extension of the set of whole numbers, and reference was made briefly to a further extension to the set of rational numbers. Among the important ideas and techniques which were studied there were the addition of integers and the idea of an additive inverse, or negative.

The present chapter revises and extends the previous work. Addition of integers is extended to include subtraction; multiplication of integers is introduced, and extended to include division; and the use of the distributive law of multiplication over addition is practised in its various forms.

As a preparation for the work in Chapter 3 on equations and inequations the rational numbers are also considered, under the operations of addition (including subtraction) and multiplication (including division). The distributive law is then studied, and applied.

For practical purposes the key ideas in this chapter are embodied in the commutative and distributive laws. A thorough grasp of these enables one to use rational numbers and variables on the set of rational numbers correctly and competently. For understanding, however, the key ideas are those of the identity element and inverses.

In addition, the identity element is *zero*, since the addition of zero has no effect, and each integer and rational number has an additive inverse called its *negative*. Thus the negative of (-3) is $(+3)$, and the negative of $(+3)$ is (-3); $(+3)+(-3) = 0$. This is explained in Book 2, and again in *Section* 2 of the present chapter, but difficulties with the minus sign may still occur since it is commonly used in three distinct ways:

(i) as a symbol for subtraction, e.g. $5-2$ ('five minus two')
(ii) as part of a numeral, e.g. (-3) ('negative three')
(iii) as part of the notation for an additive inverse, e.g. $(-a)$ ('the negative of a').

It is worth while ensuring a good understanding of the idea of an additive inverse, or negative, since it is fundamentally a very simple one, and gives insight into many of the processes of elementary mathematics.

In multiplication, the identity element is *one*, since multiplication by 1 has no effect. In the system of rational numbers (but not in the system of integers) each non-zero element has a multiplicative inverse called its *reciprocal*. In the system of whole numbers the concept of division is concerned with the ideas of quotition and partition, and remainders may be non-zero. But once the rational numbers are available, division as an operation ceases to be a difficulty and becomes simply a part of multiplication, namely multiplication by the reciprocal.

*   *   *

Section 1 revises the work of Book 2 on the addition of integers, and *Section* 2 revises and extends the operation to include subtraction of integers, so that subtracting b from a is the same as adding the negative of b to a, i.e. $a-b = a+(-b)$. In Exercises 2 and 4 some questions on addition and subtraction are set in column form. As well as giving additional practice, this will prepare the way for one of the methods given for solving systems of equations in Book 4.

Multiplication of integers is introduced in *Section* 4. Suitable examples and the use of pattern enable the pupils to discover the so-called 'rules of sign'. *Sections* 5 and 6 provide a wide variety of examples, and make further use of the commutative and associative laws. Exercise 9B need not be taken by all pupils, of course.

Section 7 introduces the distributive law for integers. It is absolutely essential for pupils to become expert at the techniques 'drilled' in this section as these will be used widely in connection with factors, equations and other aspects of the course. The illustrative worked examples give suitable guidance about methods and setting down, but it is hoped that abler pupils will take appropriate short cuts.

Section 8 treats the system of rational numbers at a reasonably elementary level. The development follows from similar topics earlier in the course. Exercise 14 leads pupils to the meaning of division, which should be taken carefully; in particular, the fact that division by zero is not possible should be emphasized. (If it did make sense to talk about dividing a non-zero number, say 5, by zero, then we should have $5 \div 0 =$ some quotient q, so that $5 = q \times 0$, i.e. $5 = 0$, which is absurd.) The note before Exercise 15 stresses that since division depends on multiplication the same 'rules of sign' apply.

The chapter ends with the simplification of algebraic fractions using the distributive law; again a 'B' exercise is considered to be appropriate here.

Summing up, this chapter tries to give pupils a fair understanding of the systems of integers and rational numbers, and to provide adequate practice in certain manipulative techniques which will be necessary in later parts of the course.

*   *   *

It may be of interest to teachers to note that by extending the number system to include the rationals we have a *field*. Further extension to include the real numbers will give no further algebraic properties. To describe the structure of the real or rational numbers more fully would necessitate a consideration of questions of *order*.

The Venn diagram illustrates the relation between the various number systems.

W = {whole numbers}; N = {natural numbers}
Z = {integers}; Q = {rational numbers}
R = {real numbers}; C = {complex numbers}

(facing page 19)

Addition of integers

33a $a+0$ **b** $-a+0$, where a is an integer.

34 Complete the statement: '0 is the ...... for the addition of integers'.

We saw in Book 2 that addition is a commutative operation and an associative operation for integers.

Commutative law: $a+b = b+a$.

For example, $4+(-6) = (-6)+4$.

Associative law: $(a+b)+c = a+(b+c)$.

For example: $(8+2)+(-4) = 8+[2+(-4)]$.

Note. When several operations have to be performed we often insert different kinds of brackets to indicate the order of working.

$9+[4+(-7)]$ is clearer than $9+(4+(-7))$, and means that (-7) has to be added to 4, and then the result has to be added to 9;

i.e. $9+[4+(-7)] = 9+(-3) = 6$

Exercise 1B

1 Illustrate on number lines $4+(-6)$ and $-6+4$. What is the sum of each?

2 Calculate $-3+(-2)$ and $-2+(-3)$, and illustrate these sums on the number line.

3 Calculate $(5+6)+(-3)$ and $5+[6+(-3)]$. What do you notice?

4 Show by calculation that $(-8+4)+(-7) = -8+[4+(-7)]$.

Assuming that the commutative and associative laws are true for the addition of all integers, calculate each of the following in the simplest possible way:

5 $(6+3)+(-3)$ **6** $(-2+3)+(-3)$ **7** $-8+(8+(-3))$
8 $-9+8+9$ **9** $-5+2+5$ **10** $15+(-12)+8$
11 $-9+(-5)+(-7)$ **12** $-8+15+(-12)$ **13** $-7+(-7)+(-7)$
14 $6x+(-3x)+(-3x)$ **15** $-y+y+(-2y)$ **16** $3z+(-4z)+5z$
17 $a+(-2a)+a$ **18** $3b+5b+(-8b)$
19 $-c+(-2c)+(-3c)$ **20** $7+(-3)+(-5)+9$
21 $-12+(-3)+7+3$ **22** $5x+(-6x)+2x+(-3x)$
23 $-4y+(-4y)+5y+(-10y)$ **24** $6a+3b+(-2a)+(-2b)$
25 $-2p+(-3q)+(-6p)+3q$

Algebra

Sometimes it is necessary to add *like terms* arranged in columns.

Example. Add the expressions in each of the following:

$$
\begin{array}{ll}
1 \quad -6x & \quad 2 \quad 2a-3b \quad\quad 3 \quad -p-2q-3r \\
2x & 4a+3b -p+3q+2r \\ \hline
-4x & 6a -2p+q-r
\end{array}
$$

Exercise 2

Add the expressions in each of the following:

1. $2a$
    $3a$

2. $5b$
    b

3. $6c$
    $-2c$

4. $8d$
    $-4d$

5. $3e$
    $-5e$

6. $2f$
    $-6f$

7. $-3g$
    $-2g$

8. $-4h$
    $-h$

9. $7x$
    $7x$

10. $7x$
    $-7x$

11. $-7x$
    $7x$

12. $-7x$
    $-7x$

13. $3x+4y$
    $x-2y$

14. $2p-3q$
    $-3p-3q$

15. $k-l$
    $k+l$

16. $-2m+3n$
    $-2m-4n$

17. $a+b+c$
    $a-b+3c$

18. $2a-3b+4c$
    $2a-3b-4c$

19. $p+q-3r$
    $-2p-2q+2r$

20. $2x-3y+4z$
    $-2x+3y-4z$

2 Subtraction of integers

All the integers, except zero, fall naturally into pairs such that the sum of the numbers in each pair is zero.

For example $1+(-1) = 0$, $\quad -2+2 = 0$, $\quad 100+(-100) = 0$.

We call each member of a pair the *negative*, or *additive inverse*, of the other. Thus the negative of 1 is -1, and the negative of -1 is 1.

Subtraction of integers

The sum of an integer a and its negative $(-a)$ is always zero:
$$a+(-a) = 0$$

We saw in Book 2 that *subtracting b* from a is the same as *adding the negative of b* to a, i.e. $a-b = a+(-b)$.

Examples

1. $6-(-5)$
    $= 6+5$
    $= 11$

2. $3-8$
    $= 3+(-8)$
    $= -5$

3. $-3x-(-2x)$
    $= -3x+2x$
    $= -x$

4. $-7x-5x$
    $= -7x+(-5x)$
    $= -12x$

Exercise 3

1 Copy and complete:

 a $9-(+5)$ b $9-(-5)$ c $-9-5$ d $-9-(-5)$
   $= 9+(-5)$ $= 9+...$ $= -9+...$ $= ...+...$
   $= ...$ $= ...$ $=$ $= ...$

Calculate

2 $5-(-3)$ 3 $6-4$ 4 $2-6$
5 $6-(-4)$ 6 $-3-3$ 7 $12-(-4)$
8 $-3-(-3)$ 9 $0-4$ 10 $8-9$
11 $-5-(-6)$ 12 $-12-(-4)$ 13 $-2-(-5)$
14 $-5-(-2)$ 15 $-7-(-8)$ 16 $3-7$
17 $-6-5$ 18 $-2-(-8)$ 19 $10-2$
20 $2-10$ 21 $-4-(-7)$ 22 $27-(-9)$
23 $-1-11$ 24 $-1+(-11)$ 25 $-8-(-8)$
26 $0-8$ 27 $0-(-8)$ 28 $8-0$ 29 $-8-0$

State whether each of the following is true or false:

30 $8-(-9) = 20-3$ 31 $9+(-3) > 6-3$
32 $7-(-4) = -4-7$ 33 $14+(-6) = 10-2$
34 $11-4 = 6-(-1)$ 35 $18-(-10) = 30-(-2)$
36 $3-7 > 10\times 0$ 37 $-2-4 < 0$

Algebra

Exercise 4

Simplify:

1. $5x-(-2x)$
2. $4y-2y$
3. $4y-y$
4. $-3z-(-2z)$
5. $2a-8a$
6. $b-3b$
7. $c-(-c)$
8. $-2d-(-2d)$
9. $6p-6p$
10. $-6p+(-6p)$
11. $-2q+2q$
12. $8r+5r$
13. $3x+5x-9x$
14. $5y+2y-7y$
15. $3z-2z-6z$
16. $4a-a-5a$
17. $-3b+b-4b$
18. $-2c-3c-4c$

Subtract the lower expression from the upper in each of the following:

19. $5a$ / $2a$
20. $2b$ / $4b$
21. $6c$ / $-3c$
22. $7d$ / $-d$
23. $-5e$ / $2e$

24. $-7f$ / $6f$
25. $-3g$ / $-4g$
26. $-5h$ / $-2h$
27. $7x$ / $7x$
28. $7x$ / $-7x$

29. $-7x$ / $7x$
30. $-7x$ / $-7x$
31. $3x+4y$ / $x-2y$
32. $2p-3q$ / $-3p-3q$

33. $k-l$ / $k+l$
34. $-2m+3n$ / $-2m-4n$
35. $a+b+c$ / $a-b+3c$
36. $2a-3b+4c$ / $2a-3b-4c$

37. $p+q-3r$ / $-2p-2q-3r$
38. $2x+3y+4z$ / $-2x-3y-4z$

3 Using the addition and subtraction of integers

Exercise 5

1. Plot the points A (1, 1), B (−2, 1), C (−2, −2), D (1, −2). What shape is ABCD? Find the coordinates of the *images* of A, B, C and D when the figure is moved in a direction:
   a. parallel to the *y*-axis until AB lies on the *x*-axis,
   b. parallel to the *x*-axis until AD lies on the *y*-axis.

2. A is the point $(a, 5)$, B is $(3, -3)$, C $(-3, -3)$, and ABC is an isosceles triangle with AB = AC. Find a.

   Write down the coordinates of B and C when the triangle is moved in a direction parallel to the *y*-axis until A coincides with the origin.

3. An aircraft flying at a height of 1500 m above sea-level lands at an airport 600 m above sea-level. What height has it lost?

4. A boy's scores in a game are 180, −100, −90 and 240. What is his total score?

5. If the temperature is −4°C and it rises 15°C, what is the new temperature?

   If the temperature *then* falls 20°C what will it be?

6. The heights of three villages A, B and C above sea-level are 100 m, −10 m and −25 m. How much higher is A than B? B than C? A than C?

Exercise 5B

1. An aircraft flying at −100 m above sea-level climbs to +300 m above sea-level. How much height has been gained?

2. A man's account at the bank is overdrawn by £80 (i.e. he owes the bank £80). How much must he deposit to have a credit of £150 in the bank?

3. The *n*th term of a sequence is given by the formula '*n*th term = $8 - 5n$'. For example, the second term is $8 - (5 \times 2) = 8 - 10 = -2$.

   Calculate the first six terms of the sequence.

Algebra

4 Repeat question *3* for the sequence for which the
$$n\text{th term} = 20 - 2n^2.$$

5 a A bird is flying north through the air at 10 metres per second, but the wind is blowing from the north at 11 metres per second. Explain what happens.

b Express both the northward speed of the bird through the air, and the *northward* speed of the wind in metres per second. Copy and complete: 'The bird is actually travelling north at . . . . . metres per second'.

6 Here is a table of times and temperatures (degrees Celsius):

Time	6 am	9 am	noon	3 pm	6 pm	9 pm	midnight
Temperature	−5	−1	2	8	5	0	−2

a What was the rise in temperature between:
(*1*) 6 am and 9 am   (*2*) 6 am and 3 pm   (*3*) 9 am and 6 pm?

b What was the fall in temperature between:
(*1*) 3 pm and 9 pm   (*2*) 6 pm and midnight?

c In which 3-hour period was the rise greatest? In which was it least?

4 Multiplication of integers

We have seen that positive integers are normally written like whole numbers since they behave like whole numbers. For example, $(+3) \times (+2) = (+6)$.

But how will we interpret $(+3) \times (-2)$, $(-3) \times (+2)$ and $(-3) \times (-2)$? The answer is determined by the need for the set of *all* integers to behave like the set of positive integers. Remember that this behaviour is completely described in terms of the commutative and associative laws of addition and multiplication, and the distributive law of multiplication over addition.

Since $2 \times 3 = 3 + 3$, we can interpret
$2 \times (-3)$ as $(-3) + (-3)$, i.e. (-6), which we write as -6.
Then $3 \times (-3) = (-3) + (-3) + (-3) = -9$
$4 \times (-3) = (-3) + (-3) + (-3) + (-3) = -12$
$5 \times (-3) = (-3) + (-3) + (-3) + (-3) + (-3) = -15$, and so on.

We define $0 \times (-3)$ to be 0, and so we have the sequence of multiples of -3: $0, -3, -6, -9, -12, -15, \ldots$

Multiplication of integers

Exercise 6

1. As shown above, calculate:
   $0 \times (-4)$, $1 \times (-4)$, $2 \times (-4)$, $3 \times (-4)$, $4 \times (-4)$, $5 \times (-4)$.

2. Calculate, or write down, the sequence of multiples:
   $0 \times (-5)$, $1 \times (-5)$, $2 \times (-5)$, $3 \times (-5)$, $4 \times (-5)$, $5 \times (-5)$.

3. Copy the table below, and use the results you have found to help you to fill in the *bottom left* part of the table.

4. Use the commutative property to help you to fill in the *top right* part of the table, e.g. assume that $(-1) \times 1 = 1 \times (-1)$.

5. Now complete the table, using the patterns that appear in the rows and columns as a guide.

   The table could be expanded as far as we please in either the positive or negative direction.

		\-5	\-4	\-3	\-2	\-1	0	1	2	3	4	5
					Second number							
	\-5	.	.	.	.	.	.	.	.	.	.	.
	\-4	.	.	.	.	.	.	.	.	.	.	.
	\-3	.	.	.	.	.	.	.	.	.	.	.
	\-2	.	.	.	.	.	.	.	.	.	.	.
First	\-1	.	.	.	.	.	.	.	.	.	.	.
number	0	.	.	.	.	.	.	0	0	0	0	0
	1	.	.	.	.	.	.	1	2	3	4	5
	2	.	.	.	.	.	.	2	4	6	8	10
	3	.	.	.	.	.	.	3	6	9	12	15
	4	.	.	.	.	.	.	4	8	12	16	20
	5	.	.	.	.	.	.	5	10	15	20	25

Note. Instead of using patterns to complete the table, we could have used the following.

To find the meaning of -2×-3
$$\begin{aligned}
& (-2 \times -3) + (-6) \\
&= (-2 \times -3) + (2 \times -3) \\
&= (-2 + 2) \times (-3), \text{ using the distributive law} \\
&= 0 \times (-3) \\
&= 0
\end{aligned}$$

It follows that -2×-3 is the additive inverse of -6, i.e. $-2 \times -3 = 6$.

In fact, for all integers a and b, $-a \times -b = ab$.

Algebra

Use your table to *write down* the following products:

6	-5×3	7	$3 \times (-5)$	8	$-2 + (-2)$
9	-1×4	10	$-4 \times (-1)$	11	$-2 \times (-3)$
12	$0 \times (-3)$	13	5×0	14	$-1 \times (-1)$
15	0×2	16	$4 \times (-4)$	17	$-3 \times (-3)$
18	-2×1	19	5×5	20	$-5 \times (-5)$
21	$4 \times (-5)$				

Of course we cannot always refer to a table, so the following results should be noted:

a The product of an integer and zero is zero.
b The product of two positive integers is positive.
c The product of two negative integers is positive.
d The product of a positive and a negative integer is negative.

Exercise 7

Use the above to calculate:

1	5×0	2	-8×0	3	8×5
4	-8×-5	5	$5 \times (-8)$	6	-5×8
7	$-10 \times (-10)$	8	$6 \times (-3)$	9	-7×2
10	$7 \times (-2)$	11	7×2	12	$-7 \times (-2)$
13	$6 \times (-9)$	14	12×1	15	$12 \times (-1)$
16	-1×100	17	$-1 \times (-100)$	18	8×7
19	-6×6	20	$-8 \times (-8)$		

Write in their simplest forms:

21	$2 \times a$	22	$-3 \times b$	23	$-4 \times (-c)$
24	$5 \times (-d)$	25	$x \times x$	26	$-x \times (-x)$
27	$y \times (-5)$	28	$-y \times (-2)$	29	$-2 \times p$
30	$-3 \times (-q)$	31	$4 \times (-r)$	32	$-s \times (-s)$

Solve the following equations, where x is a variable on the set of integers:

33	$2x = 8$	34	$2x = -8$	35	$2x = 2$
36	$2x = 0$	37	$4x = 20$	38	$4x = -4$

Using the multiplication of integers

39 $4x = 0$
40 $4x = -40$
41 $-3x = 6$
42 $-3x = 0$
43 $-3x = -3$
44 $-3x = 12$
45 $x \times (-2) = -8$
46 $x \times (-3) = 12$
47 $x \times (-1) = 0$
48 $x \times (-4) = -4$
49 $-6x = 54$
50 $-8x = -56$
51 $-9x = 9$
52 $12x = -120$

5 Using the multiplication of integers

We can use the results of the last Section in many calculations.

Example 1. If $p = -4$ and $q = -5$, calculate the values of:
    a pq **b** $6p - 2q$ **c** $(p+q)^2$ **d** $p^2 + q^2$

a pq
    $= -4 \times (-5)$
    $= 20$

b $6p - 2q$
    $= 6 \times (-4) - 2 \times (-5)$
    $= -24 + 10$
    $= -14$

c $(p+q)^2$
    $= [-4 + (-5)]^2$
    $= (-9)^2$
    $= 81$

d $p^2 + q^2$
    $= (-4)^2 + (-5)^2$
    $= 16 + 25$
    $= 41$

Example 2. If x is a variable on the set of integers, find the solution sets of:
    a $x^2 = 9$ **b** $x^3 = 1$ **c** $x^3 = -8$
Solution set is $\{3, -3\}$ Solution set is $\{1\}$ Solution set is $\{-2\}$

Exercise 8

If $a = -1, b = 2, c = 0, x = -2, y = -3$, calculate the values of the expressions in questions *1–24*:

1 ab
2 bc
3 ax
4 ay
5 bx
6 by
7 cy
8 a^2
9 b^2
10 x^2
11 y^2
12 $a^2 - b^2$
13 $x^2 - y^2$
14 $2a + x$
15 $2a - x$
16 $5x + 3y$

Algebra

17 $4x-2y$ 18 $6a+3b$ 19 $xy+3b$ 20 $ab+xy$
21 a^3 22 c^3 23 x^3 24 y^4

25 If x is replaced by an integer, which of the following can be negative?
  a x b x^2 c x^3 d x^4 e x^8 f x^9

Find the solution sets of the following, when x is a variable on the set of integers:

26 $x^2 = 16$ 27 $x^2 = 1$ 28 $x^2 = 4$ 29 $x^2 = -4$
30 $x^3 = 8$ 31 $x^3 = -27$ 32 $x^3 = -1$ 33 $x^3 = 64$
34 $x^4 = 1$ 35 $x^2 = 0$ 36 $x^2 = 100$ 37 $x^3 = -125$

Is each of the open sentences in questions **38–43** true or false for the given replacement of the variable?

38 $2-3x = 8; x = -2$ 39 $2a-12 = 0; a = -6$
40 $5x+7 = -3; x = -2$ 41 $x^2+x+1 = 0; x = -1$
42 $2y+4 > 0; y = -2$ 43 $6-3y < 0; y = 3$

44 A stone is dropped from a building, and its velocity V metres per second downwards after t seconds is given by the formula $V = 10t$.
    Calculate its velocity after:

    a 2 seconds   b 3·5 seconds

45 A ball is thrown upwards at 20 metres per second. After t seconds its upward velocity V metres per second is given by the formula $V = 20-10t$.
  a Calculate its velocity after (*1*) 1 second   (*2*) 4 seconds.
  b Explain what these results mean.

Exercise 8B

If $p = -5$, $q = -10$, $r = -1$, $s = 0$, calculate the values of the expressions in questions *1–6*:

1 pqr 2 qrs 3 $pq+qr$
4 $p^2+q^2+r^2$ 5 r^3+s^3 6 $6pr-4rs$

Find the solution sets of the equations and inequations in questions *7–18* if x is a variable on the set of integers:

7   $x^2 = 81$ 8 $x^2 = -1$ 9 $x^3 = 0$
10 $x^3 = -216$ 11 $x^3 = 1000$ 12 $(x+2)^2 = 25$

The associative law of multiplication for integers

13    $(x+7)^2 = 16$ 14 $(x-1)^2 = 4$ 15 $x^2 < 9$
16    $x^3 < 0$ 17 $x^2 \geqslant 0$ 18 $x^3 > 100$

Is each of the following (questions *19–22*) true or false for the given replacement?

19    $x^2 - x - 6 = 0;\ x = -2$ 20 $3z - 6 > 0;\ z > 2$
21    $7 - 5a < 12;\ a < -1$ 22 $5x + 10 > 0;\ x > -2$

23   When a stone is thrown vertically upwards at 15 metres per second its height h metres above the point of projection after t seconds is given by the formula $h = 15t - 5t^2$. Calculate its height after:
       a   1 second   *b*   2 seconds   *c*   3 seconds   *d*   4 seconds
Explain the answer to *d*.

24*a*   The temperature is rising steadily at 3°C per hour, and at 'zero hour' it is 20°C. What would the temperature be at 'zero hour + 2 hours'? At 'zero hour − 2 hours'?

    b   Repeat *a* for the case where the temperature is rising at −3°C per hour. What does this mean?

25   The annual profit £P from a boarding house was calculated from the formula $P = 10n - 600$, where n is the number of guest-weeks. Use the formula to calculate P when $n = 150$, 120 and 50. Explain what the results mean.

26   The length of a rectangle is $(x+3)$ metres and the breadth is $(x+2)$ metres, where x is an integer. The area is not more than 60 m². List the possible replacements for x.

27   A clock gains x minutes a day for seven days. The regulator is adjusted, and the clock then loses 1 minute a day for three days. If the total gain is 11 minutes, find x.

6   The associative law of multiplication for integers

We can calculate $-2 \times 3 \times (-4)$ as follows.

   a $(-2 \times 3) \times (-4)$ *b* $-2 \times [3 \times (-4)]$
       $= -6 \times (-4)$ $= -2 \times (-12)$
       $= 24$ $= 24$

It follows that $(-2 \times 3) \times (-4) = -2 \times [3 \times (-4)]$.

Algebra

This illustrates the associative law of multiplication for integers:
$(a \times b) \times c = a \times (b \times c)$, where a, b and c represent integers.

Exercise 9B

1 Verify by calculation that
- **a** $(-3 \times 4) \times (-5) = -3 \times [4 \times (-5)]$
- **b** $(2 \times 5) \times (-6) = 2 \times [5 \times (-6)]$
- **c** $(-4 \times -5) \times -6 = -4 \times [(-5) \times (-6)]$
- **d** $(7 \times -4) \times (-1) = 7 \times [(-4) \times (-1)]$

Use the commutative and associative laws to simplify the products in questions *2–18*:

2 $3 \times (-5) \times (-2)$ **3** $-2 \times (-7) \times 3$ **4** $-5 \times 4 \times (-2)$
5 $-7 \times 0 \times 2$ **6** $4 \times (-5) \times 1$ **7** $(-4) \times (-3) \times (-2)$
8 $2 \times (-2) \times (-5)$ **9** $(-3)^2 \times 4$ **10** $-1 \times (-5)^2$
11 $(-2)^3 \times 5$ **12** $(-3)^3 \times 2$ **13** $(-2)^2 \times (-3)^2$
14 $2^3 \times (-2)^3$ **15** $2^2 \times 5 \times (-3)$ **16** $(-1) \times (-2) \times (-3)^2$
17 $(-1) \times (-2) \times (-3) \times (-4)$ **18** $-4 \times 3 \times (-3) \times (-5)$

In questions *19–24*, $a = -1$, $b = 2$, $c = 0$, $x = -2$, $y = -3$. Evaluate:

19 $3ab - 5xy$ **20** $3xy - 2abc$ **21** $abx^2 + y^2$
22 $ab + bc + xy$ **23** $x^2y^2 - 40$ **24** $2a^2 + 3b^2 - 4x^2$

Is each of the following true or false for the replacement given?

25 $2a^2 - 3 = 5; a = -2$ **26** $5 - 4c^2 = 9; c = -1$
27 $3y^2 = 48; y = -4$ **28** $x^2 - 3x + 4 = 22; x = -3$
29 $x - 2x^3 = 51; x = -3$ **30** $8y - 2y^2 < 0; y = -4$

7 The distributive law of multiplication over addition for integers

Exercise 10

1 Copy and complete the following table:

a	b	c	$(b+c)$	$a \times (b+c)$	ab	ac	$ab+ac$
2	3	−1	2	4	6	−2	4
5	−2	3	.	.	.	.	.
−2	5	−4	.	.	.	.	.
−6	−3	−2	.	.	.	.	.

By comparing the numbers in the fifth and eighth columns do you find that in each case $a(b+c) = ab+ac$?

2 Copy and complete the following table:

a	b	c	$(b-c)$	$a \times (b-c)$	ab	ac	$ab-ac$
2	3	−4	7	14	6	−8	14
−3	5	−1	.	.	.	.	.
4	2	−5	.	.	.	.	.
−1	−5	−2	.	.	.	.	.

Do you find that $a(b-c) = ab-ac$ in each case?

Questions *1* and *2* illustrate the distributive law of multiplication over addition, which we met in Books 1 and 2. This law holds for multiplication over addition of integers, and can be stated

$$a(b+c) = ab+ac$$

Expressing a product of factors as a sum or difference of terms

  Examples

1 $-3(x-2)$ **2** $-(2p-3q)$ **3** $5(a-2b+3c)$
  $= -3x+6$ $= -1(2p-3q)$ $= 5a-10b+15c$
        $= -2p+3q$

(Remember that the product of two negative integers is positive.)

Algebra

Exercise 11

Use the distributive law to express each of the following as a sum or difference of terms:

1. $7(x-2)$
2. $-2(a+6)$
3. $-3(x-1)$
4. $5(x-1)$
5. $-6(x-y)$
6. $-8(-a-b)$
7. $-4(4-b)$
8. $-3(-x-5)$
9. $x(a-2b)$
10. $-x(-p+q)$
11. $-z(2a-b)$
12. $-x(x+y)$
13. $-3(5x-4)$
14. $-5(2m-3n)$
15. $-a(-2p-2q)$
16. $x(7-5x)$
17. $-(x-y)$
18. $-(2x-3y)$
19. $-(5-4x)$
20. $-(3a+2b)$
21. $4(a-3b+2c)$
22. $-3(a+3b-5c)$
23. $-5(1-x-x^2)$
24. $-2(2x^2-3x+4)$
25. $-(x^2-2x-3)$
26. $-(5-4x-3x^2)$

Simplifying a sum or difference of terms

Interchanging sides, the distributive law can be written:
$$ab+ac = a(b+c), \quad \text{or} \quad ba+ca = (b+c)a$$

Notice the common factor a in each case.

Examples

1. $\quad -2a-6a$
   $= (-2-6)a$
   $= [-2+(-6)]a$
   $= -8a$

2. $\quad 2x-3y+4x+5y-8x$
   $= 2x+4x-8x-3y+5y$
   $= (2+4-8)x+(-3+5)y$
   $= -2x+2y$

3. $\quad 5x-2(x-3)$
   $= 5x-2x+6$
   $= 3x+6$

Note. Each of the above examples includes the 'addition and subtraction of like terms' mentioned in Book 2.

Exercise 12

Simplify:

1. $10a-3a$
2. $3a-10a$
3. $9y-12y$
4. $-2z+6z$
5. $-3x-3x$
6. $-7q+2q$
7. $6p-10p$
8. $-5a+5a$
9. $2a+3a-4a$
10. $5b-6b-8b$
11. $10c+6c-c$

The distributive law of multiplication for integers

12    $-6k-6k-6k$
13    $-8m+5m+3m$
14    $n-4n-2n$
15    $5x-6x+2y$
16    $-3x-2x+y$
17    $8x-3y+5y$
18    $3a-4a+5b$
19    $-2a+b-2b$
20    $4a-3b-2a$
21    $5x-3x+7y-2y$
22    $6a-7a+9b-b$
23    $a+5b-2b-4a$
24    $4p+q-3q-p$
25    $5x+2(x-y)$
26    $5x-2(x-y)$
27    $6x-3(x+y)$
28    $10+3(x-2)$
29    $12-5(1-x)$
30    $4(x+2)-6x$
31    $1-2(x-1)$
32    $5x-(2x-3)$
33    $6-(2-x)$
34    $3(2x+1)-2$
35    $8x-3(2x+1)$
36    $7-4(1-2x)$
37    $3(x+y)+2(x-y)$
38    $4(x+y)-3(x-y)$
39    $5(x-y)-(x-y)$
40    $2(3x+1)-3(2x+1)$

Exercise 12B

Simplify the expressions in questions *1–18*:

1    $7x^2-3x^2+2y-9y$
2    $3a^2+6b^2-10a^2-9b^2$
3    $3x^2+2x-2x^2-3x$
4    $2(x+y)-x+3y$
5    $3(x+y)-2(x-y)$
6    $5(x-y)-(x+y)$
7    $4(a-b)+2(a+b)$
8    $-2(x+y)-(x-y)$
9    $3(-a+b)-2(a+2b)$
10    $2(x^2-x)-3(x-1)$
11    $4(y^2-1)-3(y+1)$
12    $5(1-z)-2(z-z^2)$
13    $5(x-2y)-3(2x-y)$
14    $2(5x-y)-(3x-2y)$
15    $2(x^2-2x)-3(1-2x)$
16    $5(2y^2-3y)-4(y-y^2)$
17    $(a-b)+(b-c)+(c-a)$
18    $a(b-c)+b(c-a)+c(a-b)$

19   Given $A = x+2y$, $B = y+2z$, $C = z+2x$, find the simplest forms of:
   a $A+B+C$
   b $A-B-C$
   c $2A+3B-5C$

20   Given $X = ap+bq$ and $Y = bp+aq$, show that:
   a $X+Y = p(a+b)+q(a+b) = (a+b)(p+q)$
   b $X-Y = p(a-b)+q(b-a) = (a-b)(p-q)$

Algebra

8 The set of rational numbers

You will remember that to calculate $\frac{1}{2}+\frac{1}{3}$ you had to replace $\frac{1}{2}$ by $\frac{3}{6}$, and $\frac{1}{3}$ by $\frac{2}{6}$.

$\frac{1}{2}$ and $\frac{3}{6}$ are *equivalent fractions*; there are infinitely many such fractions: $\frac{1}{2}, \frac{2}{4}, \frac{3}{6}, \frac{4}{8}, \ldots$, each of which denotes the same number called 'one half'.

The set $\{\frac{1}{2}, \frac{2}{4}, \frac{3}{6}, \ldots\}$ is a set of equivalent fractions, and the number defined by such a set is called a *rational number*.

Some rational numbers behave like integers. For example, the one which can be written $\frac{5}{1}$, or $\frac{10}{2}$, or $\frac{15}{3}$, ..., has the same properties as the integer $+5$, and is usually written '5'. The rational number which can be written $\frac{0}{1}, \frac{0}{2}, \frac{0}{3}, \ldots$ behaves like the integer 'zero', and is usually written 0.

Notation for a negative rational number

$$\frac{-2}{3} = \frac{-2\times(-1)}{3\times(-1)} = \frac{2}{-3}.$$ These are often written $-\frac{2}{3}$.

Addition and subtraction

We can add and subtract rational numbers by using our knowledge of the operations on integers, as in the following examples.

Examples

1. $\frac{2}{3}-\frac{5}{6}$
$= \frac{4}{6}-\frac{5}{6}$
$= \frac{4-5}{6}$
$= \frac{-1}{6}$
$= -\frac{1}{6}$

2. $-\frac{5}{4}-(-\frac{7}{4})$
$= -\frac{5}{4}+\frac{7}{4}$
$= \frac{-5+7}{4}$
$= \frac{2}{4}$
$= \frac{1}{2}$

3. $-\frac{1}{2}-\frac{3}{5}$
$= -\frac{5}{10}-\frac{6}{10}$
$= \frac{-5-6}{10}$
$= -\frac{11}{10}$

The set of rational numbers

Exercise 13

Write down two equivalent fractions for each of the following:

1. $\frac{3}{4}$
2. $\frac{-1}{2}$
3. $\frac{-4}{-5}$
4. $\frac{0}{6}$
5. 3

Calculate:

6. $\frac{3}{4}+\frac{1}{2}$
7. $\frac{3}{4}-\frac{1}{2}$
8. $-\frac{3}{4}+\frac{1}{2}$
9. $-\frac{3}{4}-\frac{1}{2}$
10. $\frac{1}{2}-\frac{1}{2}$
11. $\frac{4}{5}+\frac{2}{3}$
12. $\frac{4}{5}-\frac{2}{3}$
13. $-\frac{4}{5}+\frac{2}{3}$
14. $-\frac{4}{5}-\frac{2}{3}$
15. $\frac{2}{3}-\frac{2}{3}$
16. $\frac{1}{2}-\frac{1}{3}$
17. $\frac{2}{3}-\frac{3}{4}$
18. $-\frac{5}{8}-\frac{1}{2}$
19. $-\frac{1}{4}-\frac{1}{2}$
20. $-\frac{3}{4}-\frac{3}{4}$
21. $\frac{3}{5}-\frac{2}{5}$
22. $\frac{3}{5}-(-\frac{2}{5})$
23. $\frac{3}{4}-(-\frac{1}{2})$
24. $-\frac{5}{8}-(-\frac{3}{4})$
25. $-\frac{2}{3}-(-\frac{5}{3})$
26. $-\frac{1}{4}+(-\frac{3}{8})$
27. $-\frac{5}{6}+(-\frac{2}{3})$
28. $-\frac{5}{6}+\frac{1}{3}$
29. $\frac{3}{4}-0$
30. $0-\frac{3}{4}$

Multiplication and division

The set of rational numbers is denoted by Q. As in the case of integers, each rational number has a *negative*. For example, $-\frac{1}{2}$ is the negative of $\frac{1}{2}$, since $\frac{1}{2}+(-\frac{1}{2}) = 0$.

The non-zero rational numbers fall naturally into pairs in a second way; for example, $\frac{2}{3}$ and $\frac{3}{2}$, since $\frac{2}{3} \times \frac{3}{2} = 1$; $-\frac{4}{5}$ and $-\frac{5}{4}$, since $-\frac{4}{5} \times -\frac{5}{4} = 1$. Each is called the *reciprocal*, or *multiplicative inverse*, of the other.

The product of a non-zero number and its reciprocal is 1: $a \times \frac{1}{a} = 1$. In this pairing, zero is left out; zero has no reciprocal.

Exercise 14

Write down the reciprocal of each of the following, if possible:

1. $\frac{5}{6}$
2. $\frac{4}{3}$
3. $\frac{2}{1}$
4. $\frac{1}{2}$
5. $\frac{2}{3}$
6. $-\frac{2}{3}$
7. $\frac{5}{3}$
8. $-\frac{4}{7}$
9. $-\frac{1}{3}$
10. 4
11. 5
12. -2
13. $\frac{1}{4}$
14. $\frac{3}{8}$
15. 0
16. $-\frac{8}{5}$
17. -1
18. $\frac{9}{3}$

Calculate each of the following pairs, and compare the results:

19. $4 \div 2$, $4 \times \frac{1}{2}$
20. $12 \div 3$, $12 \times \frac{1}{3}$
21. $15 \div \frac{5}{2}$, $15 \times \frac{2}{5}$
22. $20 \div 2$, $20 \times \frac{1}{2}$
23. $\frac{3}{4} \div \frac{3}{4}$, $\frac{3}{4} \times \frac{4}{3}$
24. $\frac{17}{5} \div 17$, $\frac{17}{5} \times \frac{1}{17}$

Algebra

The meaning of division

What is meant by $-12 \div 4$, or $\dfrac{-12}{4}$? It is the number which, when *multiplied* by 4, gives -12, i.e. the number -3.

Division is related to *multiplication* in the same way that *subtraction* is related to *addition*.

$-8-2$ is the number which, when 2 is added, gives -8; i.e. -10.

$-8 \div 2$ is the number which, when multiplied by 2, gives -8; i.e. -4.

Just as *subtracting* an integer or a rational number is the same as *adding its negative*, so *dividing* by an integer or a rational number is the same as *multiplying by its reciprocal*.

Since zero has no reciprocal, division by zero is not possible.

Examples

1 $-12 \div 4$ *2* $\dfrac{20x}{-5x}$

$= -12 \times \tfrac{1}{4}$ $= -4$

$= \dfrac{-12}{4}$

$= -3$

Check: $-3 \times 4 = -12$ *Check:* $-4 \times (-5x) = 20x$

3 $-10 \div (-\tfrac{2}{3})$ *4* $0 \div 2$

$= -10 \times (-\tfrac{3}{2})$ $= 0 \times \tfrac{1}{2}$

$= \dfrac{-10 \times -3}{2}$ $= \dfrac{0 \times 1}{2}$

$= \tfrac{30}{2}$ $= 0$

$= 15$

Check: $15 \times (-\tfrac{2}{3}) = -10$ *Check:* $0 \times 2 = 0$

Note. It follows from the results for the multiplication of integers (see page 26) that when we divide two integers or two rational numbers:

a   if both are positive, or both are negative, the quotient is positive

b   if one is positive and the other is negative, the quotient is negative.

The set of rational numbers

Exercise 15

Calculate, where possible:

1. $-15 \div 3$
2. $12 \div (-4)$
3. $-8 \div (-2)$
4. $16 \div (-8)$
5. $-18 \div 3$
6. $20 \div (-20)$
7. $-36 \div (-4)$
8. $24 \div (-3)$
9. $7 \div (-1)$
10. $21 \div (-7)$
11. $32 \div (-4)$
12. $-36 \div (-9)$
13. $2 \div \frac{3}{2}$
14. $2 \div (-\frac{3}{2})$
15. $-4 \div \frac{1}{2}$
16. $-4 \div (-\frac{1}{3})$
17. $\dfrac{\frac{4}{5}}{\frac{4}{3}}$
18. $\dfrac{\frac{3}{8}}{-\frac{1}{8}}$
19. $\dfrac{-\frac{3}{4}}{\frac{1}{2}}$
20. $\dfrac{-\frac{3}{4}}{-\frac{3}{2}}$
21. $-\frac{7}{8} \div (-\frac{3}{4})$
22. $0 \div 5$
23. $0 \div \frac{1}{4}$
24. $2 \div 0$
25. $\dfrac{16}{-8}$
26. $\dfrac{-12}{-3}$
27. $\dfrac{-20}{4}$
28. $\dfrac{-5}{-15}$
29. $\dfrac{30x}{5x}$
30. $\dfrac{30x}{-5x}$
31. $\dfrac{-30x}{5x}$
32. $\dfrac{-30x}{-5x}$
33. $\dfrac{-10a}{5a}$
34. $\dfrac{6b}{-2b}$
35. $\dfrac{-14c}{-2c}$
36. $\dfrac{-4d}{d}$
37. $\dfrac{12x}{-12x}$
38. $\dfrac{0}{2x}$
39. $\dfrac{0}{-5y}$
40. $\dfrac{-2z}{6z}$

The distributive law for rational numbers

The distributive law is true for rational numbers, and can be used to simplify fractions.

Example. $\dfrac{-21x+7y}{7}$

$= \frac{1}{7}(-21x+7y)$

$= \frac{1}{7} \times (-21x) + (\frac{1}{7} \times 7y)$

$= -3x + y$

Alternatively $\dfrac{-21x+7y}{7}$

$= \dfrac{-21x}{7} + \dfrac{7y}{7}$

$= -3x + y$

Exercise 16B

Copy and complete:

1. $\frac{1}{2}(8+4a)$
   $= (\frac{1}{2} \times 8) + (\frac{1}{2} \times 4a)$
   $= \ldots + \ldots$

2. $\dfrac{9c+12}{3}$
   $= \frac{1}{3}(9c+12)$
   $= \ldots \ldots$

3. $\frac{2}{3}(6x-9y)$
   $= (\frac{2}{3} \times 6x) - (\frac{2}{3} \times 9y)$
   $= \ldots - \ldots$

Algebra

Simplify:

4. $\frac{1}{2}(10+4m)$
5. $\frac{1}{4}(4x+8)$
6. $\frac{3}{4}(4p+12q)$
7. $\frac{2}{3}(12a-6b)$
8. $\frac{3c+9}{3}$
9. $\frac{2x-2y}{2}$
10. $\frac{6+6a}{6}$
11. $\frac{4x+5xy}{x}$

Copy and complete:

12. $\frac{1}{2}(4c-10)$
    $= \frac{4c}{2} - \frac{10}{2}$
    $= \ldots - \ldots$

13. $\frac{-3a-9b}{3}$
    $= -\frac{3a}{3} - \frac{9b}{3}$
    $= \ldots \ldots$

14. $\frac{8p-12}{-4}$
    $= \frac{8p}{-4} - \frac{12}{-4}$
    $= \ldots \ldots$

Simplify:

15. $\frac{15t-20}{5}$
16. $\frac{-4c-6}{2}$
17. $\frac{12x+15}{-3}$
18. $\frac{8y-4}{-4}$
19. $\frac{1}{7}(14-7x)$
20. $\frac{1}{2}(10-2a)$
21. $\frac{1}{3}(-3p-6q)$
22. $\frac{1}{4}(-8+8z)$
23. $\frac{-15x-15y}{5}$
24. $\frac{2p+8q}{-2}$
25. $\frac{-3a-6b}{3}$
26. $\frac{15-5x}{-5}$
27. $\frac{x-xy}{x}$
28. $\frac{a^2-ab}{a}$
29. $\frac{7x-3xy}{-x}$
30. $\frac{-4x-12xy}{-4x}$

Given $a = 2$, $b = -1$, $c = -3$, $d = 0$, evaluate where possible:

31. $\frac{a+b}{a}$
32. $\frac{a-b}{c}$
33. $\frac{a+c}{b}$
34. $\frac{d-c}{a}$
35. $\frac{a+b+c}{c}$
36. $\frac{abcd}{2}$
37. $\frac{ab-ac}{ab}$
38. $\frac{a}{d}$

Summary

1 *Addition* is
  (i) a commutative operation: $a+b = b+a$
  (ii) an associative operation: $(a+b)+c = a+(b+c)$

$$a+0 \qquad\qquad -3y+(-7y) \qquad\qquad -\tfrac{3}{4}+\tfrac{1}{2}$$
$$= 0+a \qquad\qquad = -10y \qquad\qquad\quad = -\tfrac{3}{4}+\tfrac{2}{4}$$
$$= a \qquad\qquad\qquad\qquad\qquad\qquad\qquad\;\; = -\tfrac{1}{4}$$

2 *Subtraction.* Subtracting b from a is the same as adding the negative of b to a, i.e. $a-b = a+(-b)$

$$9a-4a \qquad\quad 9a-(-4a) \qquad\quad -9a-(-4a)$$
$$= 9a+(-4a) \quad = 9a+4a \qquad\quad = -9a+4a$$
$$= 5a \qquad\qquad = 13a \qquad\qquad\quad = -5a$$

3 *Multiplication* is
  (i) a commutative operation: $ab = ba$
  (ii) an associative operation: $(ab)c = a(bc)$

Properties. $a \times 0 = 0 \times a = 0; \quad a \times 1 = 1 \times a = a$

The product of two positive or two negative integers or rational numbers is positive; the product of a positive and negative number is negative.

$$6 \times \tfrac{2}{3} \qquad 6 \times (-\tfrac{2}{3}) \qquad -6 \times \tfrac{2}{3} \qquad -6 \times -\tfrac{2}{3}$$
$$= 4 \qquad\; = -4 \qquad\;\; = -4 \qquad\;\; = 4$$

4 *Multiplication* is distributive over addition:

$$a(b+c) = ab+ac, \quad \text{or} \quad ab+ac = a(b+c)$$
$$2(x-3) \qquad -3(x-2y) \qquad 5x-(2x+3)$$
$$= 2x-6 \qquad = -3x+6y \qquad = 5x-2x-3$$
$$\qquad\qquad\qquad\qquad\qquad\qquad = 3x-3$$

5 *Division.* If a and b are non-zero rational numbers, dividing a by b is the same as multiplying a by the reciprocal (or multiplicative inverse) of b, i.e. $a \div b = a \times \dfrac{1}{b}$

Division by zero is not possible.

$$6 \div \tfrac{2}{3} \qquad 6 \div (-\tfrac{2}{3}) \qquad -6 \div \tfrac{2}{3} \qquad -6 \div (-\tfrac{2}{3})$$
$$= 6 \times \tfrac{3}{2} \quad = 6 \times (-\tfrac{3}{2}) \quad = -6 \times \tfrac{3}{2} \quad = -6 \times (-\tfrac{3}{2})$$
$$= 9 \qquad\;\; = -9 \qquad\qquad = -9 \qquad\quad = 9$$

Equations and Inequations in One Variable

1 Equations and inequations

The equation $\quad x+5 = 9, x \in W$ . . . . (1)
and the inequation $\quad x+5 < 9, x \in W$ . . . . (2)
are *open sentences*.

Neither sentence is true or false until the variable x is replaced by a whole number.

In (1), 4 is the only replacement for x which gives a true sentence. Hence the *solution of the equation* is 4, and the *solution set* is $\{4\}$.

In (2), replacements 0, 1, 2 and 3 for x give true sentences; all other replacements from W give false sentences. The *solutions of the inequation* are therefore 0, 1, 2 and 3, and the *solution set* is $\{0, 1, 2, 3\}$.

The process of finding the solution set of an equation or inequation is called *solving* the equation or inequation.

In this chapter we investigate some principles which will help us to solve many kinds of equations and inequations.

2 Adding the same number to each side of an equation

Exercise 1
(Mainly for discussion)

1 a Is $5+3 = 8$ a true sentence?
  b Add 4 to each side. Is $(5+3)+4 = 8+4$ a true sentence?
  c Add (-12) to each side. Is $(5+3)+(-12) = 8+(-12)$ a true sentence?

Note to the Teacher on Chapter 3

It is important that pupils develop a reasonable facility at this stage in solving equations and inequations. Equations of several different types will be met in a variety of contexts later in the course, and the methods introduced here will be useful on many occasions. *This chapter aims to give an efficient technique for solving equations and inequations, based on a good understanding of the principles involved.*

Sections 2–4 deal with the solution of equations, and *Sections* 5–8 with inequations, similar methods being used in each case. *Section* 9, on equations and inequations with fractions, adds to the manipulative requirements, and *Section* 10, on problems, calls for the analysis of a situation and the setting up of a mathematical model.

In earlier books pupils were introduced to equations and inequations which were solved 'by inspection', using the addition and multiplication tables. They now meet a systematic method of solution, using clear and concise mathematical principles. Use is made of a number of familiar ideas. These include the commutative, associative and distributive laws, the addition and multiplication of positive and negative numbers, adding negatives of numbers and multiplying by reciprocals. Consideration of the set from which the variable is replaceable deepens the pupils' acquaintance with W, Z and Q.

The technique for solving equations and inequations may be summarized as follows:

Equations
(1) Addition principle: for all $a, b, c \in R$, $\quad a = b \Leftrightarrow a+c = b+c$
(2) Multiplication principle: for all $a, b, c \in R$, $(c \neq 0)$,
$$a = b \Leftrightarrow ca = cb$$

Inequations
(1) Addition principle: for all $a, b, c \in R$,
$$a > b \Leftrightarrow a+c > b+c; \quad a < b \Leftrightarrow a+c < b+c$$
(2) Multiplication principle: for all $a, b, c \in R$, $(c \neq 0)$,
$$\text{if } c > 0, \quad a > b \Leftrightarrow ca > cb; \quad a < b \Leftrightarrow ca < cb$$
$$\text{if } c < 0, \quad a > b \Leftrightarrow ca < cb; \quad a < b \Leftrightarrow ca > cb$$

These principles are intuitively evident from the introductory exercises.

The Exercises on solving equations and inequations are carefully graded in order to give plenty of practice with each of these important ideas, without having unnecessary complications which distract attention from the essence of the matter. In the case of the more advanced questions on the addition principle, it is of course better in most cases to simplify first; but there are occasions when a judicious use of the addition principle at an early stage can help considerably. Notice also that the simplest equivalent equation may appear in the form $5 = x$.

The basic feature of the chapter is finding the set of all replacements for the variables (on a defined set) which make the open sentence, i.e. the equation or inequation, into a true sentence. This provides the solution set of the equation or inequation, obtained from the *simplest equivalent* equation or inequation.

Section 1 notes that solving an equation such as $2x-6 = 0$ really entails finding its solution set, but in practice it may be preferable to write the solution as $x = 3$ (the simplest equivalent equation). The instructions in the Exercises make it clear when a solution set is required, and the worked examples give suitable guidance. The teacher is advised to refer to Book 1, pages 24–29, at this stage.

Where difficulty is experienced in using the reciprocal, a two-stage method might be used. As an illustration, the worked examples preceding Exercises 4 and 10 could be solved as follows:

$$-\tfrac{3}{8}x = 12 \qquad\qquad -4x < 20$$
$$\Leftrightarrow \tfrac{3}{8}x = -12 \qquad\qquad \Leftrightarrow 4x > -20$$
$$\Leftrightarrow \tfrac{8}{3}\times\tfrac{3}{8}x = \tfrac{8}{3}\times(-12) \qquad \Leftrightarrow \tfrac{1}{4}\times 4x > \tfrac{1}{4}\times(-20)$$
$$\Leftrightarrow x = -32 \qquad\qquad \Leftrightarrow x > -5$$

It is worth while encouraging pupils to check (perhaps mentally) some of the solutions obtained. For example, to verify that $x = 7$ is a solution of $2x+4 = 18$, L.S. $= (2\times 7)+4 = 14+4 = 18 =$ R.S.

Section 10 gives practice in the preparation of mathematical models involving equations and inequations. Here the sequence is 'Problem → mathematical model, including equation or inequation and its solution → answer to the problem.'

Adding the same number to each side of an equation

2 *a* Is $\frac{1}{2}+\frac{1}{4} = \frac{3}{4}$ a true sentence?
  b Add $(-\frac{1}{4})$ to each side. Simplify each side. Do you get a true sentence?

3 *a* Add (-5) to each side of the equation $x+5 = 9$.
  b Simplify each side of the new equation, and give its solution set.
  c Is this the solution set of the first equation?

4 *a* Add 3 to each side of the equation $x-3 = 7$.
  b Simplify each side of the new equation, and give its solution set.
  c Is this the solution set of the first equation?

The answers to questions *1* and *2* illustrate that if we add the same number to each side of an equality we obtain another true sentence.

The answers to questions *3* and *4* illustrate that if we add the same number to each side of an equation (which is an open sentence) we obtain another equation with the same solution set as the first. The two equations are said to be *equivalent*.

Example 1. Find the solution set of $x+7 = 3$, x being a variable on the set of integers.

$$x+7 = 3 \quad \ldots \ldots \ldots \quad (1)$$
$$\Leftrightarrow x+7+(-7) = 3+(-7) \quad (adding\ (-7)\ to\ each\ side)$$
$$\Leftrightarrow \quad x = -4 \quad \ldots \ldots \ldots \quad (2)$$

The solution set of (2) is $\{-4\}$, and (2) is equivalent to (1), as indicated by the symbol $\Leftrightarrow$ (*'is equivalent to'*). So the solution set of (1) is $\{-4\}$.

Example 2. Solve $5x-2 = 4x+7$, $x \in Z$.
$$5x-2 = 4x+7$$
$$\Leftrightarrow 5x-2+2 = 4x+7+2 \quad (adding\ 2\ to\ each\ side)$$
$$\Leftrightarrow \quad 5x = 4x+9$$
$$\Leftrightarrow 5x+(-4x) = 4x+(-4x)+9 \quad (adding\ (-4x)\ to\ each\ side)$$
$$\Leftrightarrow \quad x = 9$$

To solve an equation we have to find the simplest equivalent equation. To do this we arrange that variables are on one side (usually the left side), and constants are on the other side, by *adding appropriate negatives to each side* of the given equation.

Algebra

Exercise 2

By adding suitable numbers to each side, solve the following equations, the variables being on the set of integers. Give the solution in the form '$x = 11$'.

1	$x+4 = 15$	*2*	$x+7 = 19$	*3*	$y+4 = 1$
4	$z+5 = -2$	*5*	$6+p = 13$	*6*	$21+q = 75$
7	$t+(-2) = 8$	*8*	$x+(-7) = 13$	*9*	$x+(-5) = -9$
10	$2y = y+4$	*11*	$5p = 4p-10$	*12*	$3m = 8+2m$
13	$2x+3 = x+7$	*14*	$3t+1 = 2t+4$	*15*	$2y+7 = 8+y$
16	$8v+10 = 7v+5$	*17*	$3x-2 = 2x+5$	*18*	$3w-2 = 2w-5$
19	$3p-44 = 6+2p$	*20*	$5x+10 = 15+4x$	*21*	$5y-4y+7 = 19$
22	$3x+12-2x = 12$	*23*	$5y+10-4y = -10$	*24*	$4z+4 = 3z-9$

Exercise 2B

Find the solution set of each of the following equations, the variables being on the set of rational numbers. Give the solution set in the form '{1}'.

1	$x+\frac{1}{2} = \frac{3}{2}$	*2*	$y+\frac{1}{2} = \frac{3}{4}$	*3*	$t+(-\frac{1}{2}) = \frac{1}{4}$
4	$z+\frac{1}{2} = \frac{1}{4}$	*5*	$x-\frac{1}{2} = -\frac{1}{4}$	*6*	$x-\frac{3}{8} = -\frac{1}{8}$
7	$0 = \frac{3}{2}-x$	*8*	$\frac{3}{4} = 1-x$	*9*	$\frac{3}{2} = \frac{1}{2}-x$
10	$\frac{1}{2}x = 3-\frac{1}{2}x$	*11*	$3x = 2x+\frac{1}{3}$	*12*	$4x-\frac{1}{2} = 3x-\frac{1}{2}$

3 Multiplying each side of an equation by the same number

Exercise 3
(Mainly for discussion)

1 a Is $3 \times 4 = 12$ a true sentence?
  b Multiply each side by 2. Is $(3 \times 4) \times 2 = 12 \times 2$ a true sentence?
  c Is $(3 \times 4) \times \frac{1}{6} = 12 \times \frac{1}{6}$ a true sentence?
  d Is $(3 \times 4) \times (-2) = 12 \times (-2)$ a true sentence?

Multiplying each side by the same number

2 *a* Is $3 \times 7 = 21$ a true sentence?
   b Multiply each side by $\frac{1}{3}$, the reciprocal of 3.
   c Is $\frac{1}{3} \times (3 \times 7) = \frac{1}{3} \times 21$?

3 *a* Multiply each side of the equation $-3x = 18$ by $-\frac{1}{3}$, the reciprocal of -3.
   b Simplify each side of the new equation, and give its solution set.
   c Is this the solution set of the first equation?

4 *a* Multiply each side of the equation $\frac{2}{3}x = 5$ by $\frac{3}{2}$, the reciprocal of $\frac{2}{3}$.
   b Simplify each side of the new equation, and give its solution set.
   c Is this the solution set of the first equation?

The answers to questions *1* and *2* illustrate that if we multiply each side of an equality by the same number we obtain another true sentence.

The answers to questions *3* and *4* illustrate that if we multiply each side of an equation by the same number we obtain another equation with the same solution set as the first, i.e. an *equivalent* equation.

Example 1. Find the solution set of $3x = 7$, $x \in Q$.

$$3x = 7 \qquad \qquad \qquad \qquad (1)$$
$$\Leftrightarrow \tfrac{1}{3} \times 3x = \tfrac{1}{3} \times 7 \quad (\textit{multiplying each side by } \tfrac{1}{3})$$
$$\Leftrightarrow \quad x = \tfrac{7}{3} \qquad \qquad \qquad \qquad (2)$$

The solution set of (2) is $\{\tfrac{7}{3}\}$, and as (2) is equivalent to (1) the solution set of (1) is $\{\tfrac{7}{3}\}$.

Example 2. Solve $-\tfrac{3}{8}x = 12$, $x \in Q$.

$$-\tfrac{3}{8}x = 12$$
$$\Leftrightarrow (-\tfrac{8}{3}) \times (-\tfrac{3}{8}x) = (-\tfrac{8}{3}) \times 12 \quad (\textit{multiplying each side by } (-\tfrac{8}{3}))$$
$$\Leftrightarrow \quad x = -32$$

To solve equations of the kind shown above we *multiply each side by the reciprocal* of the coefficient of the variable.

Exercise 4

In questions *1–15*, give the reciprocal (or multiplicative inverse) of each number.

1 2   **2** 5   **3** $\tfrac{5}{8}$ **4** $\tfrac{1}{2}$ **5** -2

Algebra

6	−13	7	$-\frac{5}{8}$	8	$-\frac{1}{2}$	9	4	10	−4
11	1	12	−1	13	$\frac{3}{4}$	14	$-\frac{3}{4}$	15	$-\frac{6}{5}$

By multiplying each side by suitable reciprocals find the solution sets of the following equations, where $x \in Q$ (the set of rational numbers).

16 $2x = 10$	17 $5x = 30$	18 $4x = 4$	19 $3x = 2$
20 $6x = -48$	21 $7x = -3$	22 $2x = 1$	23 $5x = 0$
24 $8x = -4$	25 $5x = 3$	26 $-x = 2$	27 $-3x = 1$
28 $2x = \frac{1}{2}$	29 $3x = \frac{2}{3}$	30 $4x = \frac{8}{5}$	31 $9x = \frac{3}{4}$
32 $\frac{1}{2}x = \frac{3}{2}$	33 $\frac{2}{3}x = 8$	34 $-\frac{1}{4}x = -1$	35 $-\frac{4}{5}x = 8$
36 $\frac{x}{4} = 5$	37 $-\frac{x}{3} = -7$	38 $\frac{x}{2} = -1$	39 $\frac{x}{6} = 0$

4 Using negatives and reciprocals to solve equations

In solving equations by adding negatives and multiplying by reciprocals we arrange to have variables on one side of the equation and constants on the other side, and try to find the simplest equivalent equation.

Example. Solve $3x-4 = 32 + 7x$, $x \in Q$.

$$\begin{aligned} & & 3x-4 &= 32+7x & \\ &\Leftrightarrow & 3x-7x &= 32+4 & &\text{(since 'adding the negative' of a} \\ &\Leftrightarrow & -4x &= 36 & &\text{number is the same as 'subtracting'} \\ &\Leftrightarrow & (-\tfrac{1}{4})\times(-4x) &= (-\tfrac{1}{4})\times 36 & &\text{the number).} \\ &\Leftrightarrow & x &= -9 & & \end{aligned}$$

Exercise 5

Solve these equations, where the variables are on the set of rational numbers.

1 $2x+3 = 9$	2 $2y+5 = 6$	3 $3m+4 = 25$
4 $5t-1 = 4$	5 $2m-3 = 11$	6 $5z-4 = 3$

Using negatives and reciprocals to solve equations

7    $4n+5 = -11$ 8 $6+3x = 14$ 9 $3x+2 = 0$
10    $5x-3 = -3$ 11 $0 = 5-2y$ 12 $7 = 4-3x$
13    $2t+\frac{1}{2} = \frac{3}{4}$ 14 $5x-\frac{5}{3} = 0$ 15 $\frac{1}{3}x+2 = 5$
16    $4x+3 = 2x+15$ 17 $7y-3 = 3y+17$ 18 $4p+12 = 48-2p$
19    $6m-2 = 2m-8$ 20 $4x-8 = 6x+12$ 21 $3t+6 = 5t+7$
22    $-7z = 3z-30$ 23 $9t+5 = 15t-1$ 24 $2x+9 = 5x+7$

Example. Find the solution set of $2(x+3) = 8-3(x-4)$, $x \in Q$.

$$2(x+3) = 8-3(x-4)$$
$$\Leftrightarrow 2x+6 = 8-3x+12$$
$$\Leftrightarrow 2x+6 = 20-3x$$
$$\Leftrightarrow 2x+3x = 20-6$$
$$\Leftrightarrow 5x = 14$$
$$\Leftrightarrow x = \tfrac{14}{5}$$

The solution set is $\{\tfrac{14}{5}\}$.

Exercise 5B

Find the solution sets of these equations where the variables are on Q.

1    $3x+4 = 2(x+11)$ 2 $2x = 3(5-x)$
3    $10(y-6) = -5y$ 4 $5(z-4) = 3(8-z)$
5    $3(t-3) = 5(2t+1)$ 6 $3x-2 = 6-(8+3x)$
7    $5(6-p)+1 = p+1$ 8 $2(m+3)-1 = 8+3m$
9    $5x+17(2+3x) = 16(1+4x)$ 10 $5t-9(t-1) = 3t+5$
11    $4-(x-3) = 6-4x$ 12 $4(13-z) = 9z-(z-7)$
13    $3(5-x) = 4(3x+2)+27$ 14 $3(2x-3)-2(1-x)-(x+1) = 0$

5 Adding the same number to each side of an inequation

Exercise 6
(Mainly for discussion)

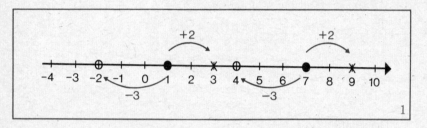

1

1 a Is $1 < 7$ a true sentence?
  b Add 2 to each side. Is $1+2 < 7+2$ a true sentence?
  c Add (-3) to each side. Is $1+(-3) < 7+(-3)$ a true sentence?
  d How are the answers to *a*, *b* and *c* illustrated in Figure 1?

2 a Is $5+3 > 6$ a true sentence?
  b Add (-3) to each side. Is $(5+3)+(-3) > 6+(-3)$ a true sentence?

In questions *3* and *4*, x is a variable on the set $\{-4, -3, -2, -1, 0, 1, 2, ..., 10\}$.

3 a What is the solution set of the inequation $x+3 < 6$?
  b Add (-3) to each side of the inequation. Simplify each side of the new inequation, and give its solution set.
  c Is this the solution set of the first inequation?

4 a What is the solution set of the inequation $x-2 \geqslant 3$?
  b Add 2 to each side. Simplify the new inequation, and give its solution set.
  c Is this the solution set of the first inequation?

This Exercise illustrates the fact that if we add the same number to each side of an inequation we obtain an equivalent inequation. So the method of Section 2 for solving equations can also be used to solve inequations.

The use of set-builder notation

Example. Find the simplest inequation equivalent to

(i) $x+3 < 8$, $x \in Z$
$x+3 < 8$
$\Leftrightarrow x+3+(-3) < 8+(-3)$
$\Leftrightarrow \qquad x < 5$

(ii) $3x \geqslant 2x-4$, $x \in Q$.
$3x \geqslant 2x-4$
$\Leftrightarrow 3x-2x \geqslant -4$
$\Leftrightarrow \qquad x \geqslant -4$

Exercise 7

Find the simplest equivalent inequation for each of the following in the form $x > 3$, $x \leqslant 3$, etc.

1	$x+5 > 8$	*2*	$y+3 < 7$	*3*	$z+2 > -1$
4	$7+p \leqslant 12$	*5*	$9+t \geqslant 4$	*6*	$6+y < 20$
7	$m-2 < 5$	*8*	$x-4 > 10$	*9*	$v-8 < -3$
10	$9 < 3-x$	*11*	$15 > 5-y$	*12*	$-2 \geqslant 10-x$
13	$2y < y+12$	*14*	$4x \geqslant 3x+15$	*15*	$3m > 2m-1$
16	$3+x > 3$	*17*	$3x+1 > 2x+5$	*18*	$5y-4 < 4y-4$

6 The use of set-builder notation

When variables are on sets such as Z or Q it is not usually possible to *list* all the members of the solution set of an inequation. But we can *describe* the solution set in several ways.

For example, the solution set of the inequation (i) in the Worked Example at the top of this page may be described as follows:

a {..., $-1, 0, 1, 2, 3, 4$}, using dots to indicate $-2, -3, -4$, and so on.
b 'the set of integers less than 5'.
c 'the set of all x such that $x < 5$, where $x \in Z$'.
   c can be given by the set-builder notation $\{x: x < 5, x \in Z\}$, or $\{x: x < 5\}$ if set Z is assumed.

The solution set of Worked Example (ii) can be written $\{x: x \geqslant -4, x \in Q\}$.

Algebra

Example. Find the solution set of $7x-2 > 6x+8$, $x \in Z$.

$$7x-2 > 6x+8$$
$$\Leftrightarrow 7x-6x > 8+2$$
$$\Leftrightarrow \quad x > 10$$

The solution set is $\{x: x > 10, x \in Z\}$

Exercise 8

1 Given $A = \{1, 2, 3, 4, 5, 6\}$, $B = \{2, 4, 6, 8, 10\}$, and $C = \{3, 5, 7, 9, 11, 13, 15\}$, list each of these sets:
- a $\{x: x > 3, x \in A\}$
- b $\{x: x < 6, x \in B\}$
- c $\{y: 5 < y < 13, y \in C\}$
- d $\{z: z+3 = 7, z \in A\}$
- e $\{t: t+1 < 10, t \in B\}$
- f $\{p: p \text{ is prime, and } p \in C\}$

Use set-builder notation to give the solution sets of the following inequations, where the variables are on the set of rational numbers.

2 $y+4 < 30$ **3** $p+12 < 9$ **4** $x+10 > -32$
5 $24+z \geqslant 15$ **6** $m-14 > 25$ **7** $t-4 \leqslant 4$
8 $t+4 \geqslant 4$ **9** $y+\frac{1}{2} < \frac{5}{2}$ **10** $p-\frac{4}{5} > \frac{3}{5}$
11 $5x > 4x+9$ **12** $2x+5 > x+12$ **13** $4y+9 < 3y-1$
14 $2m-\frac{3}{2} < m-\frac{1}{2}$ **15** $3n-1 \leqslant 8+2n$ **16** $2x+7 > 1+x$

---

7 Multiplying each side of an inequation by the same number

Exercise 9

(Mainly for discussion)

1 State whether each of the following sentences is true or false.
- a
  - (1) $12 < 18$
  - (2) $12 \times 2 < 18 \times 2$
  - (3) $12 \times \frac{1}{6} < 18 \times \frac{1}{6}$
  - (4) $12 \times 0 < 18 \times 0$
  - (5) $12 \times (-1) < 18 \times (-1)$
  - (6) $12 \times (-1) > 18 \times (-1)$
  - (7) $12 \times (-\frac{1}{3}) < 18 \times (-\frac{1}{3})$
  - (8) $12 \times (-\frac{1}{3}) > 18 \times (-\frac{1}{3})$
- b
  - (1) $12 > 8$
  - (2) $12 \times 3 > 8 \times 3$
  - (3) $12 \times \frac{1}{4} > 8 \times \frac{1}{4}$
  - (4) $12 \times 0 > 8 \times 0$
  - (5) $12 \times (-1) > 8 \times (-1)$
  - (6) $12 \times (-1) < 8 \times (-1)$
  - (7) $12 \times (-\frac{1}{4}) > 8 \times (-\frac{1}{4})$
  - (8) $12 \times (-\frac{1}{4}) < 8 \times (-\frac{1}{4})$

Multiplying each side of an inequation by the same number

Question *1* illustrates that:
- *a* if we multiply each side of an inequality by the same *positive* number we obtain another inequality which is true.
- *b* if we multiply each side of an inequality by the same *negative* number we obtain an inequality which is true if, and only if, the *direction* of the inequality symbol is reversed.

2 *a* Give two solutions of the inequation $x > 4$, $x \in Q$.
  b Do these solutions belong to the solution sets of:
  - (*1*) $2x > 8$ (multiplying each side by 2)
  - (*2*) $-x > -4$ (multiplying each side by -1)
  - (*3*) $-x < -4$ (multiplying each side by -1, and reversing the inequality sign)?

3 *a* Give two solutions of the inequation $x < -3$, $x \in Q$.
  b Do these solutions belong to the solution sets of:
  - (*1*) $2x < -6$ (multiplying each side by 2)
  - (*2*) $-x < 3$ (multiplying each side by -1)
  - (*3*) $-x > 3$ (multiplying each side by -1, and reversing the inequality sign)?

4 *a* Give two solutions of the inequation $-x > -10$, $x \in Q$.
  b Do these solutions belong to the solution sets of:
  - (*1*) $-2x > -20$ (multiplying each side by 2)
  - (*2*) $x > 10$ (multiplying each side by -1)
  - (*3*) $x < 10$ (multiplying each side by -1, and reversing the inequality sign)?

Questions *2–4* illustrate that:
- *a* if we multiply each side of an inequation by a *positive* number we obtain an equivalent inequation.
- *b* if we multiply each side of an inequation by a negative number we obtain an equivalent inequation if, and only if, the direction of the inequality symbol is reversed.

Example. Find the simplest equivalent inequation for $-4x < 20$, $x \in Q$.

$$-4x < 20$$
$$\Leftrightarrow -\tfrac{1}{4} \times (-4x) > -\tfrac{1}{4} \times 20 \text{ (Note the reversal of the inequality}$$
$$\Leftrightarrow \qquad x > -5 \qquad \text{symbol.)}$$

Note. The solution set is $\{x : x > -5, x \in Q\}$.

Algebra

Exercise 10

Find the simplest equivalent inequation (in the form $x > 2$, $x \leqslant 2$, etc.) for each of the following:

1	$\frac{1}{2}x > 1$	2	$\frac{1}{2}y < 3$	3	$\frac{1}{3}z \leqslant \frac{1}{3}$	4	$\frac{1}{4}x > 0$
5	$2t < 10$	6	$3x > 12$	7	$7x \geqslant 7$	8	$4t < 3$
9	$\frac{2}{3}w < 1$	10	$\frac{3}{4}x < 9$	11	$2x > -6$	12	$3y \leqslant -2$
13	$-4x < 12$	14	$-\frac{1}{5}z > 1$	15	$-2x < 0$	16	$-x < 5$
17	$-x < -5$	18	$-5z < 6$	19	$-\frac{3}{4}x \leqslant -3$	20	$-\frac{2}{3}y > \frac{3}{2}$

Find the solution set of each of the following, the variables being on Q:

21	$\frac{1}{2}x > 5$	22	$\frac{1}{4}y < 1$	23	$\frac{2}{3}z > 2$	24	$\frac{3}{4}x \leqslant 6$
25	$2p > 18$	26	$3x \leqslant 24$	27	$2x > \frac{3}{2}$	28	$\frac{3}{4}y < \frac{1}{2}$
29	$4y \leqslant 12$	30	$\frac{1}{2}p \leqslant 10$	31	$4x > -28$	32	$9m < -108$
33	$72n < -72$	34	$-6t > 18$	35	$-3x > 21$	36	$-6y < 18$
37	$-5x > -15$	38	$8x < 0$	39	$-x \geqslant 100$	40	$-7y \geqslant 0$

8 Using negatives and reciprocals to solve inequations

The methods are similar to those used for equations in Section 4.

Example. Solve $8 > 3x-4$, $x \in Z$.

When the variables occur only on the right side of an equation or inequation it is possible to rewrite the equation or inequation as in (ii) below, or to proceed as in (iii).

(i) $\qquad 8 > 3x-4$
$\Leftrightarrow \qquad -3x > -4-8$
$\Leftrightarrow \qquad -3x > -12$
$\Leftrightarrow -\frac{1}{3} \times (-3x) < -\frac{1}{3} \times -12$
$\Leftrightarrow \qquad x < 4$

(ii) $\qquad 8 > 3x-4$
$\Leftrightarrow 3x-4 < 8$
$\Leftrightarrow \qquad 3x < 12$
$\Leftrightarrow \qquad x < 4$

Using negatives and reciprocals to solve inequations

(iii) $\quad 8 > 3x - 4$
$\Leftrightarrow 4 + 8 > 3x$
$\Leftrightarrow \quad 12 > 3x$
$\Leftrightarrow \quad\quad 4 > x$
$\Leftrightarrow \quad\quad x < 4$

The solution set is $\{x: x < 4, x \in Z\}$

Exercise 11

Find the solution set of each of the following, where the variables are on the set of integers.

1. $4x - 28 > 0$
2. $3y + 15 < 0$
3. $2x - 1 > 3$
4. $2y + 1 < 9$
5. $2t + 5 \geqslant 13$
6. $3x - 2 < 4$
7. $5m + 1 > 16$
8. $3x + 2 < 11$
9. $2x + 9 \leqslant 5$
10. $3t + 10 < 7$
11. $4z - 3 > -35$
12. $5t - 9 > -4$
13. $9x + 5 < 5$
14. $12 - 4y \leqslant 0$
15. $40 - x \geqslant x$
16. $x < 2x - 3$
17. $3y - 5 \geqslant 4y$
18. $z + 5 < 3z$

Example. Find the solution set of $15 - 8x \leqslant 2x + 30$, $x \in Q$.

$$15 - 8x \leqslant 2x + 30$$
$\Leftrightarrow \quad -8x - 2x \leqslant 30 - 15$
$\Leftrightarrow \quad\quad\quad -10x \leqslant 15$
$\Leftrightarrow -\tfrac{1}{10} \times (-10x) \geqslant -\tfrac{1}{10} \times 15$
$\Leftrightarrow \quad\quad\quad\quad x \geqslant -\tfrac{3}{2}$

The solution set is $\{x: x \geqslant -\tfrac{3}{2}, x \in Q\}$

Exercise 11B

Find the solution set of each of the following, $x \in Q$.

1. $3m + 1 < -1 + m$
2. $3x + 1 > x - 2$
3. $4p - 7 > 2p$
4. $5 - 2y < 4$
5. $x - 2 \geqslant 6 + 3x$
6. $2y + 3 < 27 - 4y$
7. $5z - 4 > 7z + 9$
8. $3p + 5 < p - 11$
9. $15 - 7x \geqslant 3x + 5$
10. $2(x + 1) > 1$
11. $3(x - 2) < -2$
12. $3(x + 1) < x + 5$
13. $4(x - 3) < x + 3$
14. $3(2x - 1) > 2(2x + 3)$
15. $2(4 - 3x) < 4(x - 5)$

Algebra

9 Equations and inequations with fractions

When there are several fractions in an equation or inequation, the best way to proceed is to form an equivalent equation or inequation without fractions. This can be done by multiplying each side by the LCM of the denominators. The resulting equation or inequation can then be solved in the usual way.

Example. Solve $\dfrac{x}{2} - 2 = 5 + \dfrac{x}{3}$, $x \in Z$.

$$\dfrac{x}{2} - 2 = 5 + \dfrac{x}{3}$$

$\Leftrightarrow 6\left(\dfrac{x}{2} - 2\right) = 6\left(5 + \dfrac{x}{3}\right)$ (multiplying each side by 6, the LCM of 2 and 3)

$\Leftrightarrow 3x - 12 = 30 + 2x$ (using the distributive law)
$\Leftrightarrow 3x - 2x = 30 + 12$
$\Leftrightarrow \quad x = 42$

Exercise 12

By first obtaining equivalent equations or inequations without fractions, find the solution sets of the following; the variables are on the set of rational numbers.

1. $\frac{1}{2}x + 3 = 9$
2. $\frac{1}{2}x - 5 = 10$
3. $\frac{1}{2}y - 1 = \frac{1}{4}$
4. $\frac{2}{3}x + 1 < \frac{1}{2}$
5. $\frac{1}{4}x + 3 > \frac{3}{4}$
6. $\frac{3}{4}x + \frac{1}{2} < \frac{1}{2}x$
7. $\dfrac{2m}{3} - \dfrac{1}{3} > \dfrac{3m}{4}$
8. $\dfrac{y}{4} - \dfrac{y}{5} = 1$
9. $\dfrac{2t}{3} - \dfrac{t}{2} > 1$
10. $\dfrac{z}{2} = \dfrac{z}{7} - 10$
11. $\dfrac{s}{3} + 2 > \dfrac{s}{2}$
12. $\dfrac{2x}{3} + \dfrac{1}{3} = 5$
13. $\frac{1}{3}(2x - 3) \leqslant 5$
14. $\frac{1}{4}(3m - 1) = 8$
15. $\frac{2}{3}(y + 1) > \frac{3}{4}$
16. $\frac{1}{3}(5y - 1) > \frac{1}{2}(2y + 1)$
17. $\frac{1}{3}(t + 2) < t + \frac{1}{2}t$
18. $\frac{1}{3}(p + 2) - \frac{1}{4}(p - 2) = 1$

Equations and inequations with fractions

Example. Find the solution set of $\dfrac{2x-3}{3} - \dfrac{x-3}{2} > 1\tfrac{1}{5}$, $x \in Q$.

$$\dfrac{2x-3}{3} - \dfrac{x-3}{2} > 1\tfrac{1}{5}$$

$\Leftrightarrow \quad \tfrac{1}{3}(2x-3) - \tfrac{1}{2}(x-3) > \tfrac{6}{5}$
$\Leftrightarrow \quad \tfrac{30}{3}(2x-3) - \tfrac{30}{2}(x-3) > 30 \times \tfrac{6}{5}$ (*multiplying each side by* 30, *the*
$\Leftrightarrow \quad 10(2x-3) - 15(x-3) > 36$ LCM *of* 3, 2 *and* 5)
$\Leftrightarrow \quad 20x - 30 - 15x + 45 > 36$
$\Leftrightarrow \quad\quad\quad\quad 20x - 15x > 36 + 30 - 45$
$\Leftrightarrow \quad\quad\quad\quad\quad\quad\; 5x > 21$
$\Leftrightarrow \quad\quad\quad\quad\quad\quad\;\; x > \tfrac{21}{5}$

The solution set is $\{x: x > \tfrac{21}{5}, x \in Q\}$

Exercise 12B

Find the solution sets of the following; the variables are on Q.

1. $\tfrac{1}{3}(x+2) + \tfrac{1}{2}(x-1) = 1$ 2. $\tfrac{1}{2}(x+5) - \tfrac{1}{4}(x+1) > 3$

3. $\dfrac{n-3}{4} + \dfrac{n-2}{3} < 5$ 4. $\dfrac{n-4}{2} - \dfrac{n-5}{5} = -1$

5. $\dfrac{t-2}{4} - \dfrac{t-4}{6} \geqslant \tfrac{2}{3}$ 6. $\dfrac{y+4}{4} - \dfrac{3y-9}{7} < \tfrac{1}{2}$

7. $\dfrac{m}{2} - \dfrac{m}{3} = \dfrac{1-m}{6}$ 8. $\dfrac{2m}{3} - \dfrac{3m-1}{2} > 0$

9. $x \neq 0$ in this question

   a $\dfrac{8}{x} = \dfrac{2}{3}$ (Multiply each side by $3x$)

   b $\dfrac{3}{x} = \dfrac{1}{2}$ c $\dfrac{2}{x} = \dfrac{3}{5}$ d $\dfrac{1}{2} = -\dfrac{3}{x}$ e $-\dfrac{5}{x} = -\dfrac{2}{3}$

10a $\dfrac{n}{8+n} = 9, n \neq -8$ b $\dfrac{x+2}{x-1} < 2, x \neq 1$

11a $\dfrac{1}{x} - \dfrac{1}{2} = \dfrac{1}{3}, x \neq 0$ b $\dfrac{1}{2} = \dfrac{1}{v} - \dfrac{1}{4}, v \neq 0$

12   $\dfrac{4x+2}{3} + \dfrac{2x+1}{2} = \dfrac{6x+3}{4}$ 13 $\dfrac{1-x}{3} - \dfrac{1-2x}{4} + \dfrac{1-3x}{5} \geqslant 0$

Algebra

10 Applications to problems

When solving a practical problem which requires the use of mathematics the first step is to set up a *mathematical model* of the problem. To do this we translate the data into one or more equations or inequations whose solutions are then used to solve the given problem.

Example. The length of a rectangle is twice its breadth. Given that the perimeter of the rectangle is 51 cm, find the length and breadth.

Mathematical model. (A diagram is often helpful.)

Let the breadth be x cm.
Then the length is $2x$ cm.
So $2x+2x+x+x = 51$
$\Leftrightarrow \qquad 6x = 51$
$\Leftrightarrow \qquad x = 8 \cdot 5$

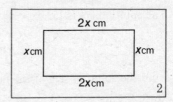

The answer to the problem is: Breadth of rectangle = 8·5 cm, length = 17 cm.

Note that x in the mathematical model is replaceable only by a positive number, in this case 8·5, and *not* by a quantity such as 8·5 cm. In other words, $x = 8 \cdot 5$, *not* 8·5 cm.

Exercise 13

1. I think of a whole number. I double it, and then add 13. The result is 37. Can you work out mentally what the number is? Now make a mathematical model using x for the number, and solve the equation. What is the number?

2. Repeat question *1* if the number is first multiplied by 4, then 3 is subtracted from the product, giving an answer of 17.

3. The length of a rectangle is $(x+3)$ mm, and the breadth is x mm.
   a. Write down the perimeter, and express it in its simplest form.
   b. If the perimeter is 30 mm, find x.
   c. What is the length and breadth of the rectangle?

4. The length of a rectangle is $(2x-5)$ cm, and the breadth is $(x+1)$ cm.
   a. Write down the perimeter, and express it in its simplest form.

Applications to problems

b Given that the perimeter is 46 cm, find x.
c What are the length, breadth and area of the rectangle?

5 The sides of a triangle are y cm, $(y+2)$ cm and $(y-3)$ cm long.

a Write down the perimeter, and simplify it.
b If the perimeter is 23 cm, find y.
c State the length of each side of the triangle.

6 The angles of a triangle are $90°$, $(x+15)°$ and $(x-15)°$.

a Find the sum of the angles in terms of x, in its simplest form.
b Form an equation in x, and solve it. What size is each angle?

7 Repeat question **6** for a triangle with angles of $4x°$, $(3x+1)°$ and $(2x-1)°$.

8 In Figure 3, the distance along the curve ABC is greater than the distance AOC. If ABC is $(3x+5)$ mm long, and AOC is $(x+12)$ mm long form an inequation in x. Find the simplest equivalent inequation.

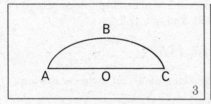

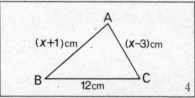

9 In Figure 4, $AB + AC > BC$. Using this information, write down an inequation in x, and find the simplest equivalent inequation.

10 One diagonal of a rectangle is marked $(8x+4)$ cm, and the other diagonal is marked $(4x+8)$ cm. Find x, and the length of each diagonal.

11 One diagonal of a kite is marked $(2x-3)$ metres, and the other is marked $(x+7)$ metres. If the first diagonal is longer than the second, what can you find out about x?

12 The length of a rectangle is $(2x+3)$ cm, and the breadth is 5 cm. Given that the area is 50 cm², find x.

13 Repeat question **12** for a rectangle with length $(4x+1)$ cm, breadth 8 cm and area 104 cm².

14 The equal sides of an isosceles triangle are each twice the length of the third side. The perimeter is 30 mm. Taking x mm for the length

Algebra

of the third side form an equation, and hence find the lengths of all three sides.

15 P is the solution set of $8x-3 > 5x-9$, and K is the solution set of $7x-2 < 3(x+6)$, $x \in Z$.
   a Find P and K, using set-builder notation to describe them.
   b List $P \cap K$.

16 3 and 4 are consecutive integers; so are 16 and 17. Given that the sum of two consecutive integers is 57, and taking n and $(n+1)$ to represent the integers, form an equation, and hence find the numbers.

17 Repeat question *16* if the sum of the two integers is -77.

18a Three consecutive integers have a sum of 75. Form an equation, solve it, and hence find the integers.
   b Is it possible for the sum of three consecutive integers to be 100?

19a The sum of three consecutive *even* numbers is 108. Find them.
   b The sum of three consecutive *odd* numbers is 63. Find them.

20 A ruler costs twice as much as a pencil. Six rulers and fifteen pencils cost £1·08. Find the cost of a pencil, and of a ruler.

Exercise 13B

1 a Find in its simplest form the perimeter of a rectangle measuring $(2a+5)$ metres by $(2a-1)$ metres.
   b Write down the perimeter of a square of side x metres.
   c If the perimeters of the square and rectangle are equal, form an equation, and express x in terms of a.
   d Which figure has the greater area when $a = 3$?

2 One packet of food costs three times as much as another. Together they cost 36 pence. Taking x pence as the cost of the cheaper packet, form an equation, solve it for x, and state the cost of each packet.

3 Plums cost twice as much per kg as apples. 3 kg of apples and 2 kg of plums cost 70 p.
   Set up a mathematical model using x pence as the price of 1 kg of apples and $2x$ pence as the cost of 1 kg of plums. Hence find the price of each.

4 The length of a rectangle is $(2x-3)$ cm, and the breadth is 5 cm. The area is less than 45 cm².
   Write down *two* inequations concerning the given information, and find the solution set of each, $x \in Q$.

Applications to problems

5 A player scores 3 points for each correct answer in a game, and pays a penalty in points for each incorrect answer. One player got 20 correct out of 60 answers, which gave him a score of -20.

Set up a model with a score of x points for each incorrect answer, and hence find this penalty.

6 The length of a rectangle is three times its breadth. If the length were 6 metres less and the breadth 6 metres more, the rectangle would be a square.

Make a mathematical model, and solve the resulting equation to find the breadth of the original rectangle. What is the area of the square?

7 In a class of 30 pupils, 19 play football, 17 play cricket, and 3 do not play either game. Suppose x pupils play *both* games, as shown in the Venn diagram in Figure 5.

 a Copy the diagram and mark, in terms of x, the number of pupils who play football only, and the number who play cricket only.
 b Form an equation for x, and solve it to find the number of pupils who play *both* games.

8 Repeat question 7 for a class of 36 pupils of whom 27 play hockey, 14 play tennis and 6 do not play either game.

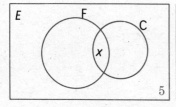
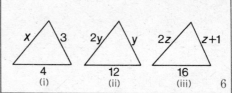

9 The 'triangle inequality' states that the sum of the lengths of any two sides of a triangle is greater than the length of the third side. Use this information to write down two inequations for each of the triangles shown in Figure 6. Where possible, give the simplest equivalent inequation.

10a How far have you travelled on a bus journey of $\frac{1}{2}$ hour at an average speed of $(2x-10)$ km/h, and on a train journey of 2 hours at $(2x+10)$ km/h?

 b Form an equation and find x if you have covered a total distance of 140 km.

Algebra

11 A motorist made a journey of 146 km. He drove for 2 hours in daylight, then for $1\frac{1}{2}$ hours in darkness at an average speed reduced by 10 km/h from the daylight speed.

  a Using x km/h for his daylight speed, write down expressions for the distances covered in daylight and in darkness.

  b Form an equation in x and solve it. Hence find: *(1)* his speed in daylight *(2)* how far he drove in the dark.

12 The total time taken by a motorist to drive 92 km was 2 hours. He averaged 54 km/h on the open road and 22 km/h in built-up areas. Make a mathematical model, using t hours for the time in built-up areas, and use it to find the time taken to pass through these areas.

Summary

1 *Addition principle*

The same number may be added to each side of an equation or inequation to give an equivalent equation or inequation, i.e. one which has the same solution set.

$$x+3 = 5, x \in W \qquad\qquad x-2 < -5, x \in Z$$
$$\Leftrightarrow x+3+(-3) = 5+(-3) \qquad \Leftrightarrow x-2+2 < -5+2$$
$$\Leftrightarrow \qquad x = 2 \qquad\qquad \Leftrightarrow \qquad x < -3$$

2 *Multiplication principle*

a Each side of an *equation* may be multiplied by the same non-zero number to give an equivalent equation.

b Each side of an *inequation* may be multiplied by the same *positive* number to give an equivalent inequation; but if the multiplier is a *negative* number the inequality symbols must be reversed.

$$5x = 3, x \in Q \qquad\qquad 5x > 3, x \in Q$$
$$\Leftrightarrow \tfrac{1}{5} \times 5x = \tfrac{1}{5} \times 3 \qquad \Leftrightarrow \tfrac{1}{5} \times 5x > \tfrac{1}{5} \times 3$$
$$\Leftrightarrow \qquad x = \tfrac{3}{5} \qquad\qquad \Leftrightarrow \qquad x > \tfrac{3}{5}$$

$$-5x > 3, x \in Q$$
$$\Leftrightarrow -\tfrac{1}{5} \times (-5x) < -\tfrac{1}{5} \times 3$$
$$\Leftrightarrow \qquad x < -\tfrac{3}{5}$$

3 *Solving equations and inequations*

To solve an equation or inequation we have to find the simplest equivalent equation or inequation.

To do this we arrange that variables are on one side (usually the left side) and constants are on the other side, by *adding an appropriate negative to each side and multiplying each side by the appropriate reciprocal*.

$$4x+7 = x-5, x \in Z \qquad\qquad 2x-3 > 5x-12, x \in Z$$
$$\Leftrightarrow 4x-x = -7-5 \qquad \Leftrightarrow \qquad 2x-5x > 3-12$$
$$\Leftrightarrow \qquad 3x = -12 \qquad \Leftrightarrow \qquad -3x > -9$$
$$\qquad x = -4 \qquad \Leftrightarrow -\tfrac{1}{3} \times -3x < -\tfrac{1}{3} \times (-9)$$
Solution set is $\{-4\}$ $\qquad\qquad \Leftrightarrow \qquad x < 3$
$$\qquad\qquad\qquad\qquad \text{Solution set is } \{x: x < 3, x \in Z\}$$

Revision Exercises

Revision Exercise on Chapter 1
Relations, Mappings and Graphs

Revision Exercise 1

1. A = {traffic light signals} = {red, red–amber, green, amber} and B = {stop, go}. Draw an arrow diagram to show the relation *means* from set A to set B.

2. In the Brown family, Mr Brown is older than Mrs Brown. Their children are George who is the eldest child, Mae, and Ann who is the youngest child. Draw an arrow diagram to show the relation *is older than* on the Brown family.

3. $P = \{-2, -1, 0, 1, 3\}$ and $Q = \{-3, 0, 2, 4\}$. Show by an arrow diagram the relation *is greater than* from set P to set Q.

4. From the set $X = \{-1, 0, 1, 2, 4\}$, write down a set of ordered pairs such that in each pair:

    a. the first number is equal to the second
    b. the second number is the square of the first
    c. the first number is two more than the second.

5. a. The relation described by x *has remainder* 2 *when divided by* 3 is defined on the set $S = \{1, 2, 3, ..., 20\}$. Write down the set A of replacements of x which satisfy this relation.

    b. Repeat *a* for the relations described by:

    (1) x *has remainder* 1 *when divided by* 3. Name this set B.
    (2) x *has remainder* 0 *when divided by* 3. Name this set C.

    c. Write down the set of all the elements in A, B and C. What do you notice?

6. a. A relation R on the set $T = \{0, 2, 3, 4, 6\}$ is described by the open

Revision Exercises on Chapter 1

sentence *x exactly divides y*. Show R as a set of ordered pairs by finding the solution set of the open sentence.

b Draw a Cartesian graph of R.

7 *a* Each of the following figures is an arrow diagram of a relation from a set A to a set B. Decide which of these relations are mappings.

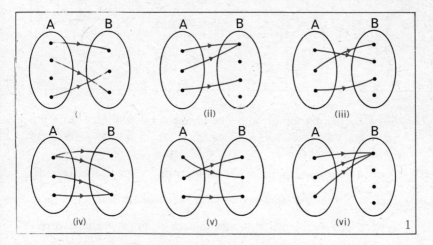

b Which mapping is a one-to-one correspondence?

8 *a* A = {Mr Smith, Mr Brown, Mr Reed}. The set of children
B = {Joe Smith, Tom Smith, Betty Smith, Jean Brown}.
Using an arrow diagram, illustrate the relation *is the parent of* from A to B. Is this relation a mapping?

b In the same way, illustrate the relation *is the child of* from B to A. Is this relation a mapping?

9 *n* is a variable on the set {0, 1, 2, 3, 4, 5, 6}. Using a Cartesian graph, illustrate the mapping described by

$$\left. \begin{array}{l} n \to 1, \text{ if } n \text{ is odd} \\ n \to 2, \text{ if } n \text{ is even} \end{array} \right\}$$

10*a* List the set of ordered pairs of the mapping *f* given by $x \to x+4$ from the set $\{-3, -2, -1, 0, 1, 2, 3\}$ to the set Z of integers.

b Draw a Cartesian graph of *f*.

c Given that the ordered pairs $(a, 25)$, $(b, -10)$ and $(c, 6\frac{1}{4})$ belong to the mapping $x \to x+4$, and *x* is a variable on the set of rational numbers, find *a*, *b*, and *c*.

Algebra

11 Figure 2 shows four relations.

a Which of these relations are mappings?

b Which mapping is a one-to-one correspondence?

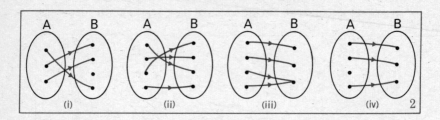

12 Show by pairing the obvious one-to-one correspondence from the set of common fractions $\{\frac{1}{2}, \frac{4}{5}, \frac{3}{20}, 5\frac{1}{4}\}$ to the set of decimal fractions $\{0.15, 5.25, 0.50, 0.80\}$.

13 A one-to-one correspondence is set up from {common fractions} to {points in the Cartesian plane} such that to each fraction $\frac{y}{x}$ there corresponds the point (x, y), and to each point (x, y) there corresponds the fraction $\frac{y}{x}$, the variables being on the set of natural numbers.

Draw a Cartesian graph to show the set of points which corresponds to each of the following sets of fractions:

a $\{\frac{1}{2}, \frac{2}{4}, \frac{3}{6}, \frac{4}{8}, \frac{5}{10}, \frac{6}{12}\}$

b $\{\frac{7}{1}, \frac{4}{3}, \frac{1}{5}, \frac{4}{7}, \frac{7}{9}, \frac{4}{11}\}$

14 Which of the following relations are mappings?

a $\{(0, 0), (2, 1), (4, 2), (6, 3), (2, -1), (4, -2), (6, -3)\}$

b $\{(0, 0), (2, 1), (4, 2), (6, 3), (8, 2), (10, 1), (12, 0)\}$

c $\{(1, a), (2, a), (3, a), (4, a), (5, a)\}$

d $\{(a, 1), (a, 2), (a, 3), (a, 4), (a, 5)\}$

15a Make a list of the ordered pairs of the mapping
$x \rightarrow \frac{1}{2}(x-4)$ from $\{-4, -2, 0, 2, 4, 6, 8, 10\}$ to the set of integers.

b Draw a Cartesian graph of the mapping.

c Using colour, show the addition to the graph you would make to illustrate the mapping $x \rightarrow \frac{1}{2}(x-4)$ on the set of *all* positive and negative numbers.

16a List the set of ordered pairs of the mapping $x \to \dfrac{12}{1+x}$ from the set $\{0, \tfrac{1}{2}, 1, 2, 3, 5, 7, 11\}$ to the set of positive numbers.

b Draw a graph of the mapping $x \to \dfrac{12}{1+x}$ on the set of positive numbers and zero.

17 Janet, Sheena, Freda and Margaret like to play badminton together but cannot all be free to play on every evening. Janet is unable to play on Tuesdays, Wednesdays and Saturdays. Sheena is free to play on Wednesdays, Thursdays and Saturdays. Freda has to stay at home on Mondays and Thursdays. Margaret can play on Mondays, Tuesdays and Fridays. None of them play on Sundays.

a On what evenings can Janet and Margaret play?
b When can Freda and Sheena play?
c On what evenings can Janet, Freda and Margaret play together?

Revision Exercises on Chapter 2
Operations on integers and rational numbers

Revision Exercise 2A

1 Calculate:

a $8+5$	**b** $8+(-5)$	**c** $-8+5$	**d** $-8+(-5)$
e $7c+4c$	**f** $7c+(-4c)$	**g** $-7c+4c$	**h** $-7c+(-4c)$

2 a Taking sea-level as zero, a point A has a height $+254$ metres and a point B has a height -35 metres. How many metres higher is A than B?

b If the temperature of a liquid changes from $-10°C$ to $-3°C$, what is the rise in temperature?

c In banking, if a pay-in is denoted by $+$ and a withdrawal by $-$, find the final state of the following account:

$$+£4·75 - £3·85 + £1·20 - £2·60$$

3 Add:

a $p+q$	**b** $2x-3y$	**c** $2a+5b$	**d** $-u-v$
$p-q$	$x-3y$	$-2a-2b$	$-2u+2v$

Algebra

4   Simplify:
- a $8-3$
- b $2-12$
- c $-5-9$
- d $-4-(-10)$
- e $4a-4a$
- f $p-3p$
- g $-3x-4x$
- h $3m-(-2m)$

5   Simplify:
- a $3c+4c+2d$
- b $5m+4m-9m$
- c $3x+5x+3y-5y$

6   Subtract the lower expression from the upper in each of the following:

- a $2x+3y$
    $x-y$
- b $3a-3b$
    $-a-2b$
- c $-p-q-r$
    $p-q+r$
- d $5x-y+2z$
    $5x+y-3z$

7   In a test, 2 marks are given for a correct answer, no marks for no answer, and -2 for a wrong answer. The test has 30 questions. What are the best and worst possible scores?

    Find the marks gained by each of the following pupils:

Pupil	Correct Answers	No Answers	Wrong Answers	Marks
Jan Brown	23	2	5	...
Jim Smith	18	8	4	...

8   Simplify:
- a 5×7
- b $7\times(-8)$
- c -9×4
- d $-4\times(-5)$
- e $-2\times p$
- f $-a\times(-a)$
- g $-a\times(-2)$
- h $-x\times 5x$

9   Solve the following equations, where x is a variable on Z:
- a $2x = 30$
- b $2x = -14$
- c $3x = 0$
- d $-3x = 15$
- e $x\times(-3) = -12$
- f $2x\times(-1) = 0$
- g $-8x = -8$
- h $-5x = 0$

10   Simplify:
- a $-2(a+3b)$
- b $5(2a-3b)$
- c $-3(4x-5)$
- d $5p-3(p-q)$
- e $10+2(x-7)$
- f $-5(1-a)-9$

11   Calculate:
- a $\frac{3}{4}-\frac{7}{8}$
- b $-\frac{2}{3}-\frac{1}{6}$
- c $-\frac{1}{2}-(-\frac{5}{8})$
- d $\frac{2}{5}-\frac{2}{3}$
- e $-20\div(-4)$
- f $-12x\div 3x$
- g $\frac{2}{3}\div(-\frac{1}{6})$
- h $-\frac{3}{4}\div(-\frac{5}{8})$

Revision Exercise 2B

1   What is the sum of:
- a $7a, -3a, -5a, 9a$
- b $-6k, -2k, 3k, 5k$
- c $-5n, -2n, 4n, -3n$
- d $4m, -10m, 5m, -3m$?

Revision Exercises on Chapter 2

2 State whether each of the following is true or false:
 - a $\quad 14-(-9) = 40-17$
 - b $\quad -5-3 > -4$
 - c $\quad -10 < 0$
 - d $\quad 8-(-3)-2 = 5+(-2)-(-2)$

3 In each of the following give three more possible terms in the sequence:
 - a $\quad 4, -1, -6, \ldots \ldots$
 - b $\quad 3, -6, 12, \ldots \ldots$
 - c $\quad -1, -4, -8, \ldots \ldots$
 - d $\quad 1, -2, 6, -24, \ldots \ldots$

4 Add *and* subtract the lower from the upper expression in each of the following:

 - a $\quad \begin{array}{l} x-y+z \\ x+y-z \end{array}$
 - b $\quad \begin{array}{l} 2a+3b-4c \\ -2a-b-c \end{array}$
 - c $\quad \begin{array}{l} p+5q+9r \\ 2p+5q-r \end{array}$
 - d $\quad \begin{array}{l} x^2+2x+1 \\ -x^2-4x+3 \end{array}$

5 Calculate:
 - a $\quad -14 \times (-9)$
 - b $\quad -6 \times (-5) \times (-4)$
 - c $\quad (-3)^3$
 - d $\quad (-0.5)^2$

6 If $p = 2$, $q = -3$, and $r = -4$, find the values of:
 - a $\quad p+q+r$
 - b $\quad pqr$
 - c $\quad pq+qr+rp$
 - d $\quad \dfrac{p-q}{q-r}$
 - e $\quad \dfrac{8q}{pr}$
 - f $\quad 3p+2q-4r$

7 Simplify:
 - a $\quad 2(-5u+3v)$
 - b $\quad 4-(4-3c)$
 - c $\quad 3(x+3)-9$
 - d $\quad 4(2a-3b)-3(4a-4b)$
 - e $\quad 2(1-x^2)-(x^2-2)$
 - f $\quad -\tfrac{1}{3}(6p-21q)$
 - g $\quad \dfrac{-8a+4b}{2}$
 - h $\quad \dfrac{9a-24b}{-3}$

8 The terms of a sequence are given by the formula $23-8x$, where x is a variable on the set $\{1, 2, 3, 4, \ldots, 20\}$.
 - a Find the first five terms of the sequence and state the last term.
 - b Does -113 belong to this sequence? If so, what is the corresponding replacement for x?

9 A rectangle is $(2x+1)$ metres long and $(2y-3)$ metres broad, where x and y are variables on the set of integers. If its length must be less than 12 metres and its breadth must not be greater than 5 metres, find all possible replacements for x and for y.

What is the area of the largest rectangle possible?

10 Given that t is a variable on the set $\{-2, -1, 0, 1, 2, 3, 4\}$, find the possible values of t^2-2t-8.

Hence find the solution set of the equation $t^2-2t-8 = 0$.

Algebra

11 Calculate:
 a $\frac{2}{7}-\frac{3}{5}$ b $-\frac{3}{4}-(-\frac{5}{6})$ c $-\frac{8}{9}\times\frac{3}{4}$ d $\frac{7}{8}\div(-\frac{5}{4})$

12 $p = 2a+3b$ and $b = 5-a$. Find a formula for p in terms of a. If $p < 10$, what can you say about a?

13 If $a \circ b = ab+2(a+b)$, calculate:
 a $3 \circ 2$ b $-\frac{1}{2} \circ 3$ c $1 \circ x$
 Find x such that $1 \circ x = 41$.

14 A warehouse gives a discount of 5 pence in the £.
 a Find the discount, in £s, on an article costing £p.
 b Obtain the net cost (reduced cost) of an article costing £p, in simplest form.
 c Find the net cost of buying 10 such articles.

Revision Exercises on Chapter 3
Equations and Inequations in One Variable

Revision Exercise 3A

1 Find the solution sets of the following equations, $x \in Z$.
 a $3x-2 = 40$ b $5x+7 = -23$ c $4x+1 = -3$
 d $5x-4 = 3x+10$ e $7-9x = 79$ f $3(x-3) = x+7$

2 Draw a number line showing the integers from -5 to 10 inclusive.
 a Show the numbers 3 and 7 on the number line by dots A and B respectively. This illustrates:
$$7 > 3 \Leftrightarrow \text{B is to the right of A} \qquad . \qquad . \qquad . \quad (1)$$
 b *Add* 2 to each number 3 and 7. Show these sums by crosses at A′ and B′ respectively on the number line. Write down a statement similar to (1).
 c *Add* (-6) to each number 3 and 7. Repeat *b* showing the sums by circles A″ and B″ respectively.
 d What do you notice about the lines AB, A′B′ and A″B″?

3 a Draw a number line to illustrate the inequality $3 < 5$, using question 2 as a guide.
 b Now show on the number line the effect of multiplying each number 3 and 5 by: (*1*) 2 (*2*) -1 (*3*) -2. Write down an inequality for each.

Revision Exercises on Chapter 3

4 Give the solution sets of the following, $x \in Q$.
- **a** $3x = 4$
- **b** $\frac{1}{4}m > 3$
- **c** $2y \geqslant \frac{1}{3}$
- **d** $\frac{2}{3}x = \frac{1}{2}$
- **e** $-3x < 18$
- **f** $-5x \geqslant -60$
- **g** $-5t = -2$
- **h** $-4p > 0$

5 Solve the following inequations, $x \in Z$.
- **a** $2x - 1 > 5$
- **b** $3x + 7 > 25$
- **c** $2(x+9) - 3 \leqslant 5$
- **d** $3x + 5 < 2x + 7$
- **e** $3x - 11 < x - 5$
- **f** $8x + 7 \leqslant 5x + 10$

6 State whether each of the following is true or false, the variables being on Q.
- **a** If $x + 4 > 8$, then $x > 4$
- **b** If $2y - 3 \leqslant 4 - y$, then $y = \frac{7}{3}$
- **c** If $7 - 5z > 3z - 9$, then $z < 2$
- **d** If $\frac{1}{2}a + 2 \geqslant \frac{1}{4}a + 3$, then $a \geqslant 4$

7 Solve the following; the variables are on Q.
- **a** $3(m - 4) = 2m + 7$
- **b** $8x + 12 < 5x + 6$
- **c** $6 - 2y > y + 5$
- **d** $\frac{1}{2}x - \frac{1}{3}x = \frac{9}{4}$
- **e** $\frac{2}{3}t + 1 \geqslant \frac{3}{2}t$
- **f** $\frac{1}{3}(6t + 9) < -15$

8 Twice a certain number added to 15 is 2 less than 27. Can you work out mentally what the number is?
Set up a mathematical model using n for the number and solve the equation. Do you still get the same answer?

9 a If k is an even number, what are the next two greater even numbers?
b Set up a mathematical model and use it to find three consecutive even numbers whose sum is 78.

10 The altimeter of an aircraft is set at zero when it is on Tehran airfield, which is well above sea-level. On take-off, the aircraft climbs for $(x + 34)$ metres and later loses height for $3x$ metres. The altimeter now reads -100 metres. Find x.

Revision Exercise 3B

1 Solve the following equations, where the variables are on Z.
- **a** $3(2c - 4) = 4c + 14$
- **b** $5(a + 3) = 2(a - 4) + 17$
- **c** $8 - 3(y + 7) = y + 5$
- **d** $15 - 6(p + 2) = 3 - 5p$
- **e** $2 + 7(2 - z) = 3z - 24$
- **f** $2(5 - 3x) + 11 = 3(2x - 5)$

2 Solve the following inequations. The variables are on Q.
- **a** $4 - x > 2$
- **b** $7 < -1 - x$
- **c** $y + 1 > y$
- **d** $7y + \frac{3}{4} < 4y + \frac{2}{3}$
- **e** $\frac{3}{2} - m < 3m + \frac{2}{3}$
- **f** $\frac{t}{3} + 1 \leqslant 2$

Algebra

3 State whether each of the following is true or false, where the variables are on the set of rational numbers:

   a If $x + \frac{1}{2} > \frac{1}{4}$, then $x > -\frac{1}{4}$ *b* $-2y > 6 \Leftrightarrow y > -3$
   c If $x > -1$ and also $x < -1$, then $x = -1$.
   d If $x < -2$ and $x \geq -4$, then $-4 \leq x < -2$.

4 A clock starts at the correct time, but gains y minutes a day for 10 days. If it is put back 5 minutes, how many minutes fast is it? The regulator is then adjusted and it loses $2y$ minutes a day for three days. If at the end of the three days it is:

   a 2 minutes fast   *b* 3 minutes slow, find y in each case.

5 Solve the following. The variables are on Q.

   a $\frac{1}{3}(a+2) > 2 + \frac{a}{2}$ *b* $\frac{x+1}{3} + \frac{x+2}{2} = 3$

   c $\frac{2y+1}{5} - \frac{y-1}{3} < \frac{2}{3}$ *d* $5 \cdot 9x - 2 \cdot 1 = 3 \cdot 4x - 1 \cdot 85$

   e $\frac{a-2}{4} - \frac{2}{3} \geq \frac{a-4}{6}$ *f* $-3y \leq 7y + \frac{1}{3}$

6 $p < q < r$ means $p < q$ and also $q < r$. Solve, $x \in Q$:

   a $-3 < 2x - 7 < 13$ *b* $-2 < \frac{x}{3} + 1 \leq 2$
   c $-7 < 5 - 2x < 1$ *d* $-2 < \frac{1}{3}(x-4) \leq -1$

7 In this question, $x \in Z$. A is the solution set of $3x + 14 > 5$ and B is the solution set of $4 - 5x > -21$. List the members of $A \cap B$ and describe also $A \cap B$ using set-builder notation.

8 A boy cycled 37 km. For part of the time he averaged 8 km/h and for the rest of the time he averaged 10 km/h. He took 4 hours altogether. Using t hours for his time at 8 km/h, set up an equation and use it to answer these questions:

   a How long did he cycle at 8 km/h?
   b How far did he cycle at 8 km/h?

9 Repeat question **8** using x km for the distance cycled at 8 km/h, and use the solution of your equation to answer the same questions. Which model is preferable?

10 A skeleton model of a rectangular box (cuboid) with two square ends is to be made of wire. The length is to be 10 cm more than the

Revision Exercises on Chapter 3

breadth. Using x cm for the breadth, find an expression in x for the total length of wire required. (A sketch will help.)

If not more than 100 cm of suitable wire is available, find the simplest inequation in x and hence show that the volume cannot exceed 375 cm³.

11 In a league competition, 3 points are awarded for a win, 2 points for a draw, and no points for losing. At the end of the season, a team had played 24 games and lost 7 of them, scoring 45 points in all. How many games did it win?

12a A container is holding 14 litres of a 25% antifreeze solution (i.e. 25% of the 14 litres is antifreeze and the rest is water). How many litres of water are in the mixture?

 b x litres of pure antifreeze are now added to the solution to bring the strength up to 30%. Write down, in terms of x, the percentage of water in the new mixture.

 c Combine your two answers to form an equation, and hence find x.

Geometry

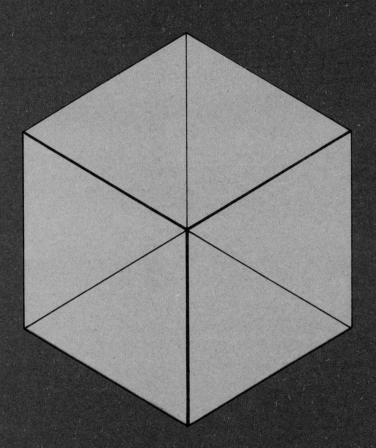

Geometry

Note to the Teacher on Chapter 1

The content of this chapter exemplifies the spirit of the course in geometry, building as it clearly does on earlier *foundations*, maintaining a number of previously established *methods*, and producing a wide variety of *results* to be exploited in later chapters. The *foundations* include the work on symmetry about a point and a line, angles, and coordinates in Book 1, and the study of the rectangle and of triangles in Book 2. The *methods* include the following.

Using 'tracing' paper, what result is *suggested*?
Using 'squared' paper, what properties are *apparent*?
Using 'certain' definitions, what can you *deduce*?
Using 'coordinates', what answers can you *work out*?

The *results* are made explicit in the Summary at the end of the chapter.

The fact that the pupils are studying **transformation geometry** or **motion geometry** in this course comes to the fore in this chapter; the opportunity to use tracings and other aids should not be missed. Points and lines and shapes must be made to move in order to suggest vividly and clearly the ideas which are crystallized in the definitions and properties in *Section* 3 and the use of coordinates in *Section* 4. All the transformations the pupils will study, reflection (Book 3), translation (Book 4), dilatation (Book 5) and rotation (Book 6) are concerned with motion, the movement or transformation of all points in the plane; physical aids and models at this early stage clarify the ideas for the pupils and provide the necessary background for the abstraction that will follow.

Section 1 explores briefly some aspects of turning shapes over and turning them round, and introduces the necessary terminology of bilateral symmetry and rotational symmetry, as well as invariant points. Rotational symmetry will be considered in more detail in the chapter on 'Rotation' in Book 6. In every rotation where a shape is conserved there is a *centre of rotation*; in the case of a figure which is conserved under a *half turn* about a point, that point is called the *centre of symmetry*. In this section, and in Chapter 2 ('The parallelogram') we are concerned only with half turns and corresponding centres of symmetry. Note also that under a half turn about O (or under a reflection in O), $P \leftrightarrow P'$ such that $PO = OP'$.

Section 2 uses these ideas to investigate properties associated with symmetry about a line. Some teachers will wish to put more emphasis on folding; this technique has largely been avoided in the writing since it may impede the development of the idea that reflection applies to the whole plane, and not just to half of it. One or two suggestions are made in the Project at the end of the section, however.

Section 3 approaches reflection in a line in a more formal way. Definitions are given concerning the image of a point in a line and the image of a line under reflection in a line, and from these several deductions are made. These are reinforced by the use of coordinates in *Section* 4, where some emphasis is put on the fact that reflection is a transformation of all points of the plane. The results

$$P(a, b) \leftrightarrow P'(a, -b)$$
and
$$P(a, b) \leftrightarrow P''(-a, b)$$

will be useful at many later stages in the course. The teacher will have to decide whether to cover the section following Exercise 5, and if so how formal to make the treatment; calculations can always be made in individual questions by reference to a diagram, but the '$a+p = 2h$' property may be of interest. Vector methods will be used at a later stage to investigate this work again.

Sections 5 and 7 emphasize the bilateral symmetries of the rhombus and kite. This approach follows smoothly from the definitions in terms of isosceles triangles. The sequential development

'*rectangle (and square)—isosceles (and equilateral) triangle— rhombus—kite*'

is worth mentioning here.

The constructions in *Section* 6 are based on the properties of the rhombus. They provide further practice in accurate drawing with instruments, and introduce certain lines associated with triangles, which will be useful later.

By the end of this chapter it should be realized that all the properties of the rectangle, square, rhombus and kite are consequences of the bilateral and rotational symmetries of these figures. Behind these, of course, lies an abstract group structure which some pupils will study later in their careers.

Reflection

1 Symmetry

1

We have all seen a picture like Figure 1 which shows the hills and houses and trees reflected in the water.

We often see our own *reflection*, or *image*, in a mirror. This chapter studies ideas about reflection in Geometry.

Exercise 1

1 a Trace each picture in Figure 2, and *turn it over* about the dotted line to check that it fits its outline again.

(*Or*, stand a small pocket mirror on the dotted line, and compare the image in the mirror with the 'image' in the diagram.)

b What points stay in the same place when the picture is turned over or reflected?

Geometry

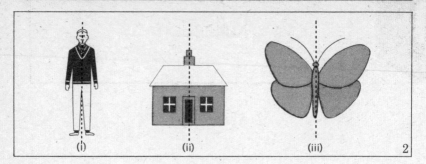

2

2 a Trace each shape in Figure 3, and *turn it round* about the coloured dot until it fits its outline again.

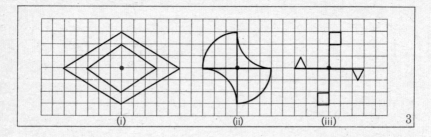

3

b Through what angles had you to turn the shapes to make them fit again?

c Which point stays in the same place during the rotation?

Points which are in the same position before and after a shape is turned over, or turned round, are called *invariant points*, or *fixed points*.

In Figure 2 the set of invariant points forms a line called the *axis* (*or line*) *of symmetry*. The shape has *bilateral symmetry* about this line.

In Figure 3 the invariant point is called the *centre of symmetry*. The shape has *half turn symmetry* about this point.

Exercise 2

1 a By tracing each diagram in Figure 4 (or by using a mirror) notice that each has at least one *axis of symmetry*. Find the position of each axis.

Symmetry

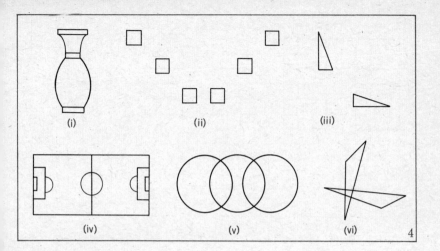

b Which diagrams have more than one axis of symmetry?

2 By tracing or studying the diagrams in Figure 5 notice that each has a *centre of symmetry*. Find the position of this centre.

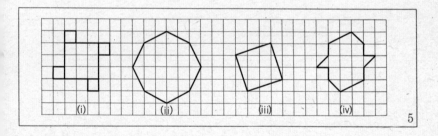

3 Copy and complete this table for the shapes in Figure 6.

Shape	a	b	c	d	e	f	g	h	i	j	k	l
Number of axes of symmetry												
Has shape a centre of symmetry?												

Geometry

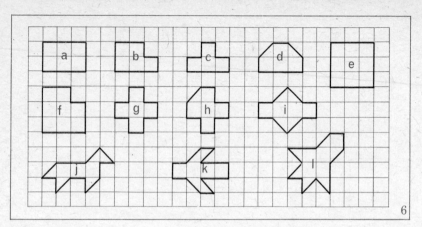

4 Given the set of capital letters {A, E, H, N, W, Z}, list subsets of letters which have:
- *a* only one axis of symmetry
- *b* two axes of symmetry
- *c* a centre of symmetry.

5 *a* Copy and complete this table:

Shape	Sketch of shape, showing centres and axes of symmetry	Number of ways of fitting its outline	Number of axes of symmetry
Rectangle			
Square			
Isosceles triangle			
Equilateral triangle			

b Comment on the numbers in the second last and last columns.

6 Make a class collection of pictures and shapes showing bilateral or half turn symmetry.

2 Symmetry about an axis

It is not difficult to draw on squared paper shapes which have bilateral symmetry. In Figure 7 the blue lines have been drawn to make the final shape symmetrical about the dotted line.

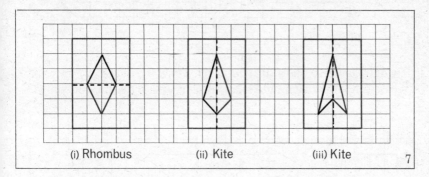

(i) Rhombus   (ii) Kite   (iii) Kite

7

Exercise 3

1 Copy the diagrams in Figure 8 on to squared paper. Complete each diagram so that the dotted line will be an axis of symmetry.

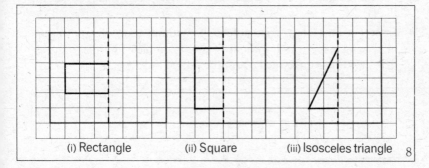

(i) Rectangle   (ii) Square   (iii) Isosceles triangle

8

Geometry

2 Repeat question *1* for Figure 9.

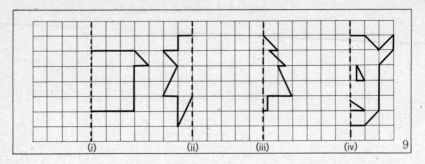

9

3 Copy Figure 10 on to squared paper. Draw the image of F in each case so that the figure has bilateral symmetry about the dotted line.

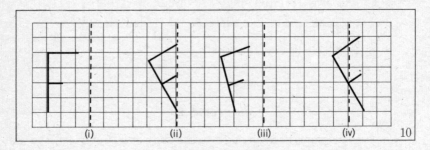

10

4 Repeat question *3* for other capital letters, for example H, M, P, N.

5 a Copy Figure 11 into your notebook. Without using squared paper or tracing paper, and without folding, complete the diagrams so that they are symmetrical about the dotted line.

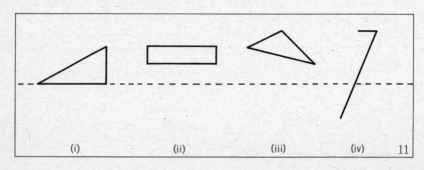

11

Reflection in a line

b Explain how to construct these images quite accurately.

Project.

Try these interesting ways of making shapes with bilateral symmetry:

(i) Make an ink-blot (in one or two colours) in the middle of a piece of paper. Fold the paper through the blot or near the blot, and press the two pieces firmly together. Unfold the paper, and study the pattern.

(ii) Fold a sheet of paper once or twice, and cut out a shape. Unfold the paper, and study the pattern.

3 *Reflection in a line*

The image of a point

To find the image of a point A under reflection in an axis XY we draw a line AM perpendicular to XY, and produce AM its own length to A′.

We say that, under reflection in XY,
A → A′ ('A goes to A′', or 'A maps to A′') and A′ → A.

More concisely, A ↔ A′

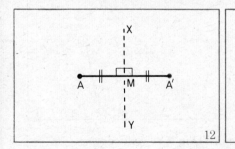
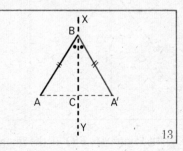

The image of a line

(i) In Figure 13 the image of AB is A′B, and is obtained as follows. Under reflection in XY, A ↔ A′ and B ↔ B.

Hence AB ↔ A′B, and AB = A′B.

Also C ↔ C, so that ∠ABC = ∠A′BC.

Geometry

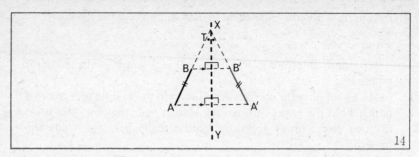

14

(ii) In the same way, in Figure 14, A'B' is the image of AB. It follows that, since AA' and BB' are both perpendicular to XY, AA' is parallel to BB'.

Also, each point on AB ↔ a point on A'B', so that if AB and A'B' meet at T, T ↔ T. It follows that AB and A'B' meet on XY, and AB and A'B' are equally inclined to XY.

Note. A shape and its image are congruent.

Exercise 4

1   In Figure 15, A' is the image of A in OP. Copy the diagram, and fill in the length of each line and the size of each angle.

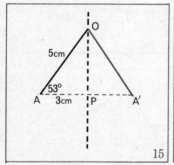

15

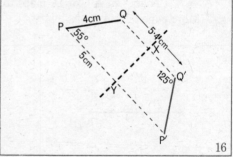

16

2   In Figure 16, P'Q' is the image of PQ in XY. Write down:
   a the lengths of PP', QX and P'Q'
   b the sizes of angles PYX, PP'Q' and PQX.

3   Draw a diagram with A on the axis XY, and B not on XY. B' is the image of B in XY. Write down the images of:
   a A   *b* B   *c* AB   *d* BB'   *e* ∠ABB'   *f* ∠BAY

Reflection in a line

4   A and B are points on XY. C is a point not on XY, and the image of C in XY is C′.
    Mark in as many pairs of equal lines and angles as you can in figure ACBC′.

5   A and B are points on the same side of XY such that AB produced meets XY at T, and $\angle ATX = 35°$. A′B′ is the image of AB in XY.
  a If AB = 7 cm and BT = 3 cm, write down the lengths of A′B′ and A′T, and the size of angle ATA′.
  b Name a line perpendicular to AA′, and a line parallel to AA′.

6   XY is an axis, and AB is a line parallel to XY. Draw the image A′B′ of AB in XY. What can you say about A′B′?

Exercise 4B

1   A and B are points on the same side of XY, and A′B′ is the image of AB in XY. AB′, A′B, AA′ and BB′ are joined. A′B cuts XY at C. Mark pairs of equal lines and angles in the diagram, and note any other special features.

2   In Figure 17, B′ is the image of B in XY, and AB′ cuts XY at P. Explain why
  a PB = PB′   b AP+PB = AB′
    Why does it follow that the shortest route from A to B via a point on XY is the route from A to P to B?

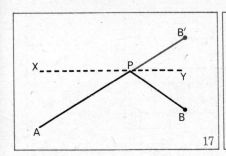

17

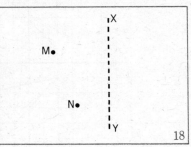

18

3   Copy Figure 18, and make a construction which gives the shortest route from M to N via a point on XY.

4   Represent the edges of a billiard table on squared paper by the lines with equations $x = 0$, $x = 6$, $y = 0$ and $y = 10$.

Geometry

A ball is struck from the position (1, 3) to reach the position (3, 7) after rebounding from one edge. Assume that the angle between the cushion and the path of the ball before impact is equal to the angle after impact.

Show by an accurate drawing four possible routes for the ball, and give the coordinates of the points where the ball strikes the edges.

5   Draw a rectangle ABCD with its diagonal AC. Show the image of the rectangle under reflection in AC, taking P as the image of B and Q as the image of D.

Why must AC, BD and PQ pass through the same point O?

Why must A, Q, B, C, P and D all lie on a circle? State the centre of this circle.

6   Under reflection in an axis XY a line AB and its image A'B' never meet. Draw a diagram to show this.

4 Coordinates

Under reflection in the x-axis, the images of A $(-2, 1)$, B $(1, 4)$, C $(3, 2)$ are A' $(-2, -1)$, B' $(1, -4)$, C' $(3, -2)$ as shown in Figure 19.

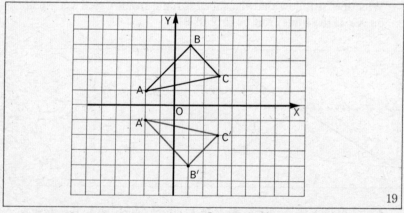

19

If X denotes the mapping '*reflection in the x-axis*', we can write: Under X, $(-2, 1) \leftrightarrow (-2, -1)$, and $\triangle ABC \leftrightarrow \triangle A'B'C'$.

A mapping of the plane (or a transformation of all points of the plane) associates with each point of the plane *one* image point of the

Coordinates

plane. The mapping given by *reflection in the x-axis* associates with the point P (a, b) its image in the x-axis, P' $(a, -b)$.

Similarly, under reflection in the y-axis, P $(a, b) \leftrightarrow$ P'' $(-a, b)$.

Exercise 5

1 a On squared paper plot the points A $(1, 6)$, B $(2, 2)$, C $(5, 4)$.
  b Plot the images of A, B and C in the x-axis, and give their coordinates.
  c Plot the images of A, B and C in the y-axis, and give their coordinates.

2 Triangle DEF is isosceles, with D $(3, 3)$, E $(-1, 2)$, F $(7, 2)$. Show triangle DEF on squared paper, and also its images under reflection in the x-axis, and in the y-axis. Give the coordinates of the vertices of the image triangles.

3 P is $(0, 5)$, Q $(1, 6)$, R $(3, 8)$, S $(5, 10)$, T $(8, 0)$.
  a Write down the images of these points under X (reflection in the x-axis) in the form $(0, 5) \leftrightarrow (0, -5)$.
  b Which of these points is invariant under this mapping?

4 Repeat question *3* for the mapping Y (reflection in the y-axis).

5 ABCD is a rectangle. A is $(4, 1)$, B $(4, 3)$, C $(0, 3)$, D $(0, 1)$. Show the images of the rectangle under reflection in the x-axis, and in the y-axis. Write down the coordinates of the vertices of the image rectangles.

6 a Find the images of the points Q $(1, 4)$, R $(2, 2)$, S $(3, 4)$, T $(4, 5)$ under reflection in the line with equation $x = 6$.
  b Repeat *a* for reflection in the line with equation $x = -2$.

7 Copy and complete this table:

Point	$(0, 0)$	$(2, 2)$	$(3, 0)$	$(0, -2)$	$(-4, -4)$
Image under X					
Image under Y					
Image under reflection in the line $x = 3$					

8 Draw a shape on squared paper. Choose an axis, and draw the image of your shape in this axis. Mark in the coordinates of all points.

* * *

Geometry

In Figure 20, P' (p, q) is the image of P (a, b) in the line $x = h$. We will find p and q in terms of a, b and h.

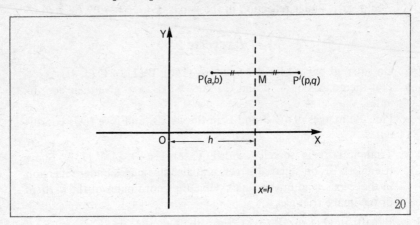

$$p = a + PP'$$
$$= a + 2PM$$
$$= a + 2(h - a)$$
$$= a + 2h - 2a$$
$$= 2h - a$$

Also $\quad q = b$

So P' is the point $(2h - a, b)$

For example, the image of P $(2, 3)$ in the line $x = 5$ is P' $(8, 3)$.

Notice that since $p = 2h - a$, $a + p = 2h$,
i.e. the sum of the x-coordinates of the point and its image is $2h$.
In the above example, $2 + p = 10$, so that $p = 8$.

Exercise 5B

1 Use the above results, or separate diagrams, to find the images of:

a P $(1, 2)$ in the line $x = 5$ *b* Q $(3, -1)$ in the line $x = 7$
c R $(6, 0)$ in the line $x = 2$ *d* S $(-3, -4)$ in the line $x = -1$

2 ABCD is a rectangle, with A $(2, 1)$, B $(5, 1)$, C $(5, 3)$, D $(2, 3)$. Find the coordinates of the vertices of the image rectangle under reflection in the line $x = 7$.

3 A is the point $(3, 2)$. Find the coordinates of its image:

a after reflection in the line $x = 5$, followed by reflection in the line $x = 10$

b after reflection in the line $x = 0$, followed by reflection in the line $y = 0$.

4 Find the images of P (1, 2), Q (3, −1), R (6, 0) and S (−3, −4) in the lines with equations: *a* $y = 0$ *b* $y = 1$.

5 Under reflection in the line $x = h$, P $(a, b) \leftrightarrow$ P'$(2h-a, b)$. What would be the coordinates of the image of P under reflection in the line $y = k$? Try to prove your result in a way similar to that following Figure 20.

6 *a* Find the images of the points K (5, 0), L (6, 6), M (4, 1), N (1, 8), P (2, −2) under reflection in the line through the origin and the point (10, 10).

b Repeat *a* for reflection in the line through the origin and the point (10, −10).

5 *The rhombus*

In Book 2 we formed the isosceles triangle from two halves of a rectangle, and found that it had one axis of symmetry as shown in Figure 21.

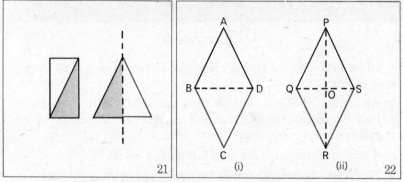

The shape ABCD formed when two congruent isosceles triangles are placed base to base as shown in Figure 22 (i) is called a *rhombus*.

It follows at once that:

a AB = BC = CD = DA (i.e. all the sides are equal)

b ∠A = ∠C, and ∠ABC = ∠ADC (i.e. opposite angles are equal)
c the diagonal BD is an axis of symmetry of the rhombus, and bisects the angles of the rhombus at B and D as shown.

Example. Prove that PR is an axis of symmetry of rhombus PQRS in Figure 22(ii).

Isosceles triangle PQS can be formed from two congruent right-angled triangles POQ and POS, so that OP is an axis of symmetry of triangle PQS, and QO = OS.

Similarly OR is an axis of symmetry of triangle QRS, so that POR is a straight line forming an axis of symmetry of the rhombus:

The diagonals of a rhombus are axes of symmetry, and bisect each other at right angles.

Note. The rhombus fits its outline in the four ways shown in Figure 23.

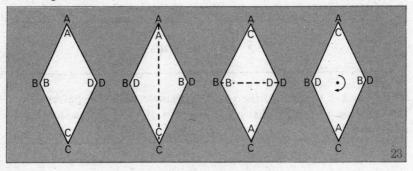

Exercise 6

1 Draw a rhombus on squared paper, and draw its axes of symmetry in colour.

2 ABCD is a rhombus, with A the point (3, 10), B (0, 5), C (3, 0). Find the coordinates of D, and of the point of intersection of the diagonals of the rhombus.

3 Repeat question *2* for A (−4, 1), B (0, 4), C (4, 1).

4 Sketch the rhombus shown in Figure 24, and mark in the length of each side and the size of each angle.

5 *a* Calculate the size of every angle in Figure 25, in which CDEF is a rhombus and ∠OCD = 40°.
  b Name a set of four equal lines, and two pairs of equal lines.

The rhombus

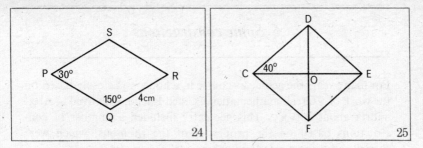

6 In rhombus ABCD, ∠BAD = 50°. Sketch the rhombus, draw in its diagonals, and calculate the sizes of all the angles in the figure.

7 a Sketch a rhombus composed of two congruent isosceles triangles PQS and RQS.
  b Prove that the sum of the angles of the rhombus is 360°.

8 State whether each of the following is true or false for every rhombus.
  a The sides are all equal.
  b The diagonals are equal.
  c The diagonals meet at right angles.
  d The diagonals bisect each other.
  e There is only one axis of symmetry.

9 P is the point $(-4, 0)$, Q $(0, -6)$, R $(4, 0)$, S $(0, 6)$. Explain why PQRS is a rhombus, and calculate its area.

10 A is the point $(2, 6)$, and C is $(8, 6)$. Draw a rhombus ABCD with a diagonal AC, and write down the coordinates of B and D. Repeat this for a different pair of points B and D. If B is the point (p, q) what can you say about p or q?

11 Calculate the areas of the rhombuses with diagonals of lengths:
  a 4 cm and 6 cm   **b** 10 cm and 16 cm.

12 Sketch a rhombus ABCD, and mark in O for the point of intersection of the diagonals. Copy and complete the following.
Under a halfturn about O, A ↔ ..., and B ↔ ...
So AB ↔ ..., and BC ↔ ...
It follows that AB is parallel to ..., and BC is parallel to ...

13 Sketch a rhombus with its diagonals, and mark in as many properties of its sides and angles as you can.

14 On squared paper draw a tiling of congruent rectangles, and from it produce a tiling of congruent rhombuses.

Geometry

6 Some constructions

For many years the geometry course in schools was closely based on the work of a Greek mathematician called Euclid, who lived in Alexandria about 300 B.C. This geometry included a number of constructions based on the properties of the rhombus, which were carried out using a ruler and compasses.

The ideas can be illustrated by means of a rhombus made from Meccano strips, with elastic diagonals. As the figure changes shape the diagonals always bisect each other at right angles, and bisect the angles of the rhombus.

(i) To construct the line from P perpendicular to AB

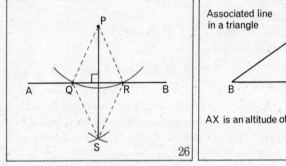
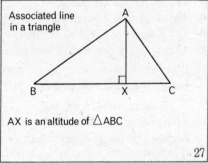

26

Associated line in a triangle

AX is an altitude of △ABC

27

With centre P, draw an arc to cut the given line AB at Q and R.
With centres Q and R, and the same radius, draw arcs to cut at S.

PQSR is a rhombus, so PS is perpendicular to AB.

Exercise 7

1. Using ruler and compasses, copy the construction shown in Figure 26, starting with a line AB about 6 cm long, and a point P above it.
2. Draw a line CD, and mark a point Q below it. Construct the line from Q perpendicular to CD.
3. *a* Draw a large triangle ABC with all its angles acute. Using ruler and

Some constructions

compasses, construct the line from A perpendicular to BC (the *altitude* from A).

b In the same way, draw the perpendiculars from B to AC, and from C to AB. If you do this accurately, you will find that the three altitudes pass through the same point, i.e. are *concurrent*.

(ii) *To bisect an angle*

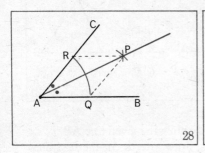

28

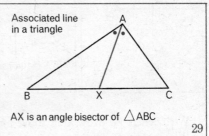

AX is an angle bisector of △ABC

29

With centre A, draw an arc to cut the arms AB and AC of ∠BAC at Q and R.
With centres Q and R, and the same radius, draw arcs to cut at P. Join AP.
AQPR is a rhombus, so AP bisects angle BAC.

Exercise 8

1 Using ruler and compasses, copy the construction shown in Figure 28 for an acute angle BAC.
2 Draw a right angle. Construct the bisector of the angle. Measure each of the two angles you have made.
3 Repeat question 2 for an obtuse angle of 120°.
4 Draw a large triangle ABC. Using ruler and compasses, draw the bisector of each angle of the triangle. If you do this accurately you will find that the bisectors of the angles are concurrent.

Geometry

(iii) *To construct the perpendicular bisector of a line*

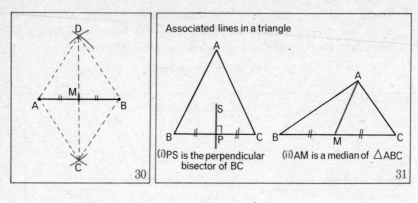

With centre A, draw arcs above and below the given line AB.
With centre B, and the same radius, draw arcs to cut the first ones at C and D.
Join CD, cutting AB at M.

ACBD is a rhombus, so (i) CD is the perpendicular bisector of AB
(ii) M is the midpoint of AB.

Exercise 9

1. Using ruler and compasses, copy the construction shown in Figure 30, starting with a line AB 6 cm long. Check that AM = MB, and $\angle AMD = 90°$.

2. Draw a large triangle ABC. Using ruler and compasses, draw the perpendicular bisector of each side. If you do this accurately, you will find that the three perpendicular bisectors are concurrent. If the point where they meet is S, you should be able to draw a circle, centre S, passing through A, B and C (the *circumcircle* of △ABC).

3. Draw a large triangle PQR. Using ruler and compasses, find the midpoints of the sides. Join P, Q and R to the midpoints of the sides opposite them. The lines you have drawn are called *medians* of the triangle, and they are concurrent lines, meeting at the *centroid* of the triangle.

7 The kite

The shape ABCD formed when two isosceles triangles with equal bases are placed base to base, as shown in Figure 32, is called a *kite*.

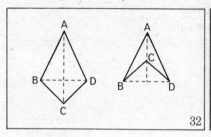
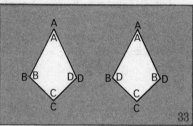

It follows that:
a AB = AD, and CB = CD
b ∠ABC = ∠ADC
c AC is an axis of symmetry (as in the Worked Example on page 86)
  Note. The kite fits its outline in the two ways shown in Figure 33.

Exercise 10

1 Draw both kinds of kite shown in Figure 32, on squared paper; letter each one ABCD. Copy and complete:
  Under reflection in AC, B ↔ ..., A ↔ ..., C ↔ ..., ∠ABC ↔ ..., △CDA ↔ ...

2 In the kite shown in Figure 34, ∠TSP = 30° and ∠PTQ = 40°. Calculate the sizes of all the other angles in the figure.

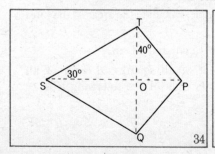
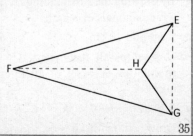

Geometry

3 In the kite in Figure 35, $\angle HEG = 35°$ and $\angle FEG = 75°$. Calculate the sizes of all the other angles in the figure.

4 a Sketch a kite composed of two isosceles triangles PQS and RQS.
  b Prove that the sum of the angles of the kite is 360°.

5 State whether each of the following is true or false for every kite:
  a The diagonals are equal.
  b The diagonals intersect at right angles.
  c The kite has two pairs of equal sides.
  d The opposite angles of the kite are equal.
  e When the diagonals are drawn there are three pairs of congruent triangles in the figure.

6 Copy and complete this table:

Shape	Sketch of shape, showing centres and axes of symmetry	Number of ways of fitting its outline	Number of axes of symmetry
Rectangle			
Rhombus			
Kite			

7 A is the point $(-2, 3)$, B is $(-4, 0)$, C $(-2, -5)$. Explain why OABC is a kite. Give the coordinates of the images of O, A, B, C under reflection in the y-axis.

8 Repeat question 7 for reflection in the x-axis.

9 a A is (8, 5), B (8, 8), C (5, 8). What shape is OABC?
  b Calculate the area of the square with diagonal OB, and the area of $\triangle OAD$, where D is (8, 0).
  c Hence calculate the area of OABC.

10 By drawing the surrounding rectangle, calculate the areas of the kites with diagonals of lengths
  a 6 cm and 10 cm   **b** 15 cm and 18 cm.

11 R = set of all rectangles, S = set of all squares, B = set of all rhombuses, K = set of all kites. Is each of the following true or false?
  a $S \subset R$ **b** $B \subset K$ **c** $R \subset B$ **d** $S \subset B$ **e** $S \subset B \subset K$

12 Draw a tiling of kites on squared paper.

8 Some problems

Exercise 11B

1. Draw a circle with an arrowhead marked on the circumference. Draw the image of this figure under reflection in a line which does not cut the circle.

2. In a rhombus PQRS, $\angle$PSR = 4$\angle$SPQ. Calculate the sizes of the angles of the rhombus.

3. XYZU is a kite in which XZ is the axis of symmetry. A is a point on XY, and B on YZ, so that AB is parallel to XZ. A' and B' are the images of A and B in XZ. What can you deduce about figure ABB'A'?

4. A rectangle has vertices O (0, 0), A (2, 0), B (2, 8), C (0, 8). Give the coordinates of the image of B under reflection in the line bisecting $\angle$OCB. Give the coordinates of the point on AB which remains invariant under the reflection.

5. The lengths of the diagonals of a kite are $2x$ cm and $2y$ cm. Show that the area of the kite is $2xy$ cm^2.

6. P is a point outside a given line XY. Using ruler and compasses construct a rhombus with one vertex at P and one diagonal on XY.

7. Two congruent triangles ABC have obtuse angles at B. They are placed on top of each other so that the angles at A coincide, and AB of one triangle lies alongside AC of the other. Show that if the sides BC cut each other at T, then AT bisects $\angle$BAC.

Geometry

Summary

1   If a figure is mapped onto itself under *reflection in a line*, the line is called an *axis of symmetry*, and the figure has *bilateral symmetry* about the axis.

2   If a figure is mapped onto itself under *a half turn about a point*, the point is called the *centre of symmetry*.
In both of the above, a figure and its *image* are congruent.

3   Under reflection in XY,

$A \leftrightarrow A'$ AA' is parallel to BB'
$B \leftrightarrow B'$ AA' and BB' are bisected at right angles
$AB \leftrightarrow A'B'$ by XY.
$AB = A'B'$ AB and A'B' meet on XY, or are
$\angle ACY = \angle A'CY$ parallel to XY.

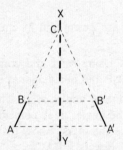 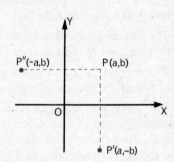

4   Under reflection in the *x*-axis, $P(a, b) \leftrightarrow P'(a, -b)$
Under reflection in the *y*-axis, $P(a, b) \leftrightarrow P''(-a, b)$

Summary

5   A *rhombus* consists of two congruent isosceles triangles base to base, and fits its outline in four ways. Also
the diagonals are axes of symmetry;
all the sides are equal;
the opposite angles are equal, and are bisected by the diagonals;
the diagonals bisect each other at right angles.

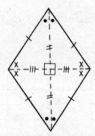 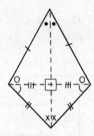

6   A *kite* consists of two isosceles triangles with equal bases, and fits its outline in two ways. Also
one diagonal is an axis of symmetry;
two pairs of sides are equal;
one pair of opposite angles is equal;
one diagonal bisects the other at right angles.

7   Lines constructed in a triangle:

Altitude   Angle bisector   Perpendicular bisector of side   Median

The Parallelogram

1 Half turns and parallel lines

Figure 1 shows a grid of congruent squares forming two sets of parallel lines.

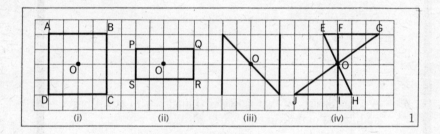

Exercise 1

1. *a* Copy and complete the following for square ABCD in Figure 1(i). Under a half turn about O,

   A ↔ ..., B ↔ ..., AB ↔ ..., AD ↔ ...

   b Name a line equal to AB, and a line equal to AD.

   c Name a line parallel to AB, and a line parallel to AD.

2. *a* Copy and complete the following for rectangle PQRS in Figure 1(ii). Under a half turn about O,

   Q ↔ ..., R ↔ ..., QP ↔ ..., QR ↔ ...

   b Name lines equal and parallel to SR and SP.

3. In how many ways does the letter N in Figure 1(iii) fit its outline? Check by means of a tracing that it fits its outline under a half turn about O.

Note to the Teacher on Chapter 2

The study of the parallelogram and its properties is developed in this chapter from two important ideas already discussed in earlier chapters, namely parallel lines and half turns; these are recalled in *Section* 1.

In *Section* 2, the parallelogram is defined in terms of a triangle and its image under a half turn. This approach leads at once to the deduction that a parallelogram is a quadrilateral with half turn symmetry from which all its properties can be deduced. Note that although rotation and translation are used in a fairly elementary fashion in this chapter, the technical terms rotation, translation and order of rotation are deliberately avoided. It is unwise to introduce too much new vocabulary into a chapter in which so many properties of parallelograms and angles associated with parallel lines have to be mastered.

In *Section* 3, to simplify the approach to a tiling of congruent parallelograms, it is first shown that a quadrilateral with two pairs of opposite sides parallel has the half turn symmetry of the definition and is therefore a parallelogram. This enables the tiling of the plane to develop naturally from two sets of equidistant parallel lines. This tiling offers opportunities of practice in half turns and translations, which will be useful in later work.

Section 4 deduces the equality of corresponding angles by translation in the parallelogram tiling and of alternate angles by half turns, a method of approach which is consistent with the general development in this chapter. It should not be difficult for pupils to find supplementary relationships of angles as well as equalities in connection with parallel lines.

A triangular tiling can easily be developed from a parallelogram tiling. The collinearity of certain diagonals in the latter is proved in Exercise 7B, question *4*. Teachers who wish to do so can construct and justify a triangular tiling from this result.

For the sake of completeness there is a short section (*Section* 5) on the area of a parallelogram. Since pupils already know the area formula for a triangle, 'area of parallelogram = twice area of triangle' seems to be the logical approach, consistent with the definition; but for those who prefer it, the direct connection between parallelogram and rectangle is suggested in a revision example.

Construction of parallelograms is not specifically mentioned but pupils already have had practice in drawing triangles. If there is sufficient information to draw half of a parallelogram, then the whole can be drawn if necessary.

(facing page 97)

Half turns and parallel lines

4 Sketch three more capital letters that fit their outlines after a half turn about their centres.

5 a Name points in Figure 1(iv) that map to E, G, I and O under a half turn about O.

b Name lines that map onto HJ, OE, FG and JG under a half turn about O.

*   *   *

Each of the shapes in Figure 1 fits its outline after a half turn about O. Each has *half turn symmetry* about O, the *centre of symmetry*.

Under reflection in O, A ↔ C, AB ↔ CD, etc. ('A goes to C, and C goes to A', or 'A maps to C, and C maps to A', or 'C is the image of A, and A is the image of C').

In Figure 2, PO is produced its own length to P′ so that PO = OP′. Under reflection in O (or under a half turn about O), P ↔ P′ (the *image* of P).

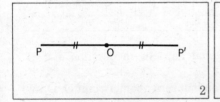
2

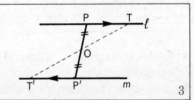
3

In Figure 3, PO = OP′, and lines *l* and *m* are parallel. Under reflection in O (or a half turn about O),

P ↔ P′ T ↔ T′, PT ↔ P′T′ and *l* ↔ *m*.

(Note that PT is parallel to its image P′T′, but in the opposite direction.)

Under reflection in a point (or a half turn about a point), a line and its image are parallel.

Exercise 2

1 a Plot the following points, and their images under reflection in the origin:

A (3, 1), B (2, 5), C (6, 0), D (0, −2), E (4, 4), O (0, 0)

b List the coordinates of the images.

Geometry

2 Without plotting the following points give the coordinates of their images in the origin:

F (2, 1),   G (−3, −3),   H (1, −4),   I (−6, 0),   J (a, b)

3 Plot the images, and hence write down the coordinates, of the following points under reflection in the point P (3, 3):

Q (5, 3),   R (3, 0),   S (5, 1),   T (−1, −1),   U (−2, 6).

4 a On squared paper draw a square ABCD, a rectangle ABCD, and an isosceles triangle ABC.
  b Draw the image of each shape under a half turn about the vertex A.
  c Mark equal lines, and parallel lines, in your diagrams.

5 Copy Figure 4 on to squared paper, and draw the image of AB under a half turn about O in each case.

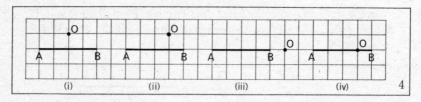

6 a Draw a diagram to illustrate what happens in question 5 if O coincides with B.
  b A line AB maps onto itself under a half turn about a point O. Draw a diagram to illustrate this.

7 In Figure 5, one line in each pair of parallel lines is the image of the other under a half turn about a point O. Copy the lines on to squared paper, and mark the position of O for each pair of lines.

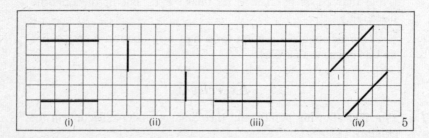

8 Copy Figure 6, in which the blue lines are parallel, and mark the centre of symmetry of each diagram.

Half turns and parallel lines

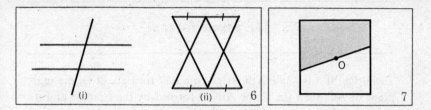

9. In Figure 7, O is the centre of the square, and the shaded area is 8 cm². Calculate the area of the square, justifying your method.

Exercise 2B

1. *a* Give the images of the following points under reflection in the origin:

   A (5, 5), B (10, −1), C (0, −12), D (*p*, *q*)

   b If $(a, b) \leftrightarrow (p, q)$, find the value of $a+b+p+q$.

2. In this question the universal set is {A, D, E, F, H, I, L, M, N, T, X}. *P* is the subset consisting of letters with half turn symmetry, and *Q* is the subset with at least one axis of symmetry. List *P*, *Q* and $P \cap Q$.

3. Draw two circles, each with radius 2 cm, centres 12 cm apart. Under reflection in point O, one circle maps onto the other. Find the position of O.

   P is a point on one of the circles. Find its image in O. Experiment with different positions of P.

4. Draw, or cut out in card, a quadrilateral. Show by successively half turning it about the midpoints of its sides that it is possible to fill the page with congruent quadrilaterals.

Geometry

2 The parallelogram

Definition. If a triangle ABC is given a half turn about O, the mid-point of AC, then the shape ABCD formed by the triangle and its image is called a parallelogram. See Figure 8.

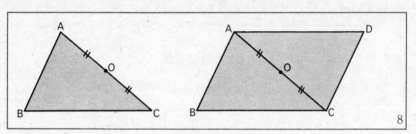

8

Exercise 3

In this Exercise we *deduce* some properties of a parallelogram. Questions *1–3* refer to Figure 8.

1   Copy and complete the following. Under a half turn about O,
- a A ↔ ..., B ↔ ..., △ABC ↔ ...
- b AB ↔ ..., so AB = ..., and AB ∥ ('is parallel to') ...
- c BC ↔ ..., so BC = ..., and BC ∥ ...
- d ∠ABC ↔ ..., so ∠ABC = ...
- e ∠ACB ↔ ..., and ∠BAC ↔ ..., so ∠BCD = ...

2 a The sum of the angles of △ABC is ...
  b The sum of the angles of parallelogram ABCD is ...
  c The sum of the angles ABC and BCD is ...

3   Parallelogram ABCD fits its outline in ... ways.

4 a On squared paper draw a triangle ABC with a right angle at B. Complete the parallelogram formed by a half turn of the triangle about the midpoint of BC.
  b Starting with the same triangle again in each case, form parallelograms by half turns of the triangle about the midpoints of AB and then AC. What special shape is this last one?

* * *

The parallelogram

Figure 9 shows a copy of the parallelogram in Figure 8, with BO and OD joined.

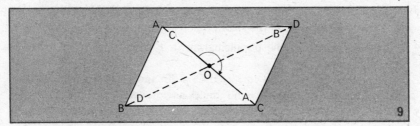

Under a half turn about O, O ↔ O.
Hence O does not change its position, so that O is *an invariant point* for the half turn.
Under a half turn about O, OB ↔ OD.
Hence B, O, D are collinear (in the same straight line), and OB = OD.
Also, OA = OC.

It follows that the diagonals of a parallelogram bisect each other. Also the parallelogram fits its outline ('is conserved') under a half turn about the point of intersection of its diagonals.

Summary of properties. A parallelogram:
  is formed by a triangle and its image under a half turn about the midpoint of a side;
  has its opposite sides equal and parallel;
  has its opposite angles equal;
  has diagonals which bisect each other;
  is conserved under a half turn about the point of intersection of the diagonals.

Exercise 4

1 Figure 10 sums up the properties of a parallelogram. Make a list of these, using the letters in the diagrams.

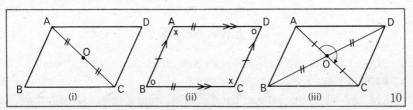

Geometry

2. In parm (short for 'parallelogram') PQRS, PQ = 8 cm, QR = 5 cm and angle P = 60°. List the sizes of the other sides and angles.

3. In parm VWXY, the diagonals intersect at O, $\angle$XOY = 100° and $\angle$OYX = 50°. Fill in the sizes of as many angles as you can.

4. *a* Make a sketch of the parallelogram in Figure 10(iii) after a half turn about O.
   b If OB = 6 cm, OC = 4 cm, $\angle$OBC = 20° and $\angle$OCB = 31°, fill in the sizes of as many lines and angles as you can.

5. The diagonals of parm EFGH intersect at M. List two pairs of parallel lines, and four pairs of equal lines, in the diagram.

6. OABC is a parallelogram. O is the origin, A is (7, 1), B (8, 6). Find the coordinates of C, and of the point of intersection of the diagonals.

7. P is (−3, −4) and Q is (1, −4). PQRS is a parallelogram with its diagonals intersecting at the origin O. Find the coordinates of R and S.

8. State whether each of the following is true or false for every parallelogram.
   a The opposite sides are parallel.
   b All the sides are equal.
   c It fits its outline in exactly two ways.
   d The opposite angles are equal.
   e All the angles are equal.
   f The sum of the angles is 180°.
   g The diagonals bisect each other.
   h It is mapped onto itself under a half turn about the point of intersection of the diagonals.

9. Repeat question *8* for every rectangle.

10. Repeat question *8* for every square.

11. *a* What additional property would you have to specify for a parallelogram to make it a rectangle?
    b What further property would make it a square?

Exercise 4B

1. PQ and RS are two straight lines which intersect at their midpoint O. Explain why PQRS is a parallelogram.

2. LM and NP are equal and parallel lines. Find a centre O such that

L ↔ P under a half turn about O. Show that this half turn must map M on to N. What kind of figure is LMPN? Why?

3 A is the point (2, 3), B (−3, −3), C (4, −3). Find the coordinates of each position of D if the following are parallelograms:

  a ABCD   *b* ACBD   *c* ACDB.

4 The diagonals of parm ABCD intersect at O. POQ and ROS are two lines meeting the sides at P, Q, R and S. Explain why PRQS is a parallelogram.

5 In parm ABCD, AB is produced to P and CD to Q so that BP = DQ. Explain why each of the following is a parallelogram:

  a APCQ   *b* BPDQ.

3 Parallelogram tiling

In Figure 11(i) ABCD is a quadrilateral with AB ∥ DC and AD ∥ BC. O is the midpoint of AC.

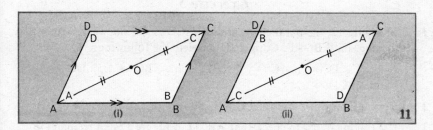

11

Under a half turn about O, as shown in Figure 11 (ii), A ↔ C, the line of AB ↔ the line of CD, and the line of CB ↔ the line of AD. It follows that B ↔ D and △ABC ↔ △CDA, and hence ABCD is a parallelogram.

A quadrilateral with its opposite sides parallel is a parallelogram.

Geometry

This is sometimes a useful definition of a parallelogram, and we make use of it in the following tiling:

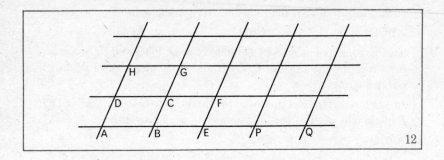

12

Figure 12 shows two sets of parallel lines, with AB = BE = EP = ..., and AD = DH = ...

It follows that ABCD is a parallelogram, and that the plane can be covered completely by congruent parallelograms, without overlapping.

Exercise 5

1. In Figure 12, the parallelogram tile ABCD is slid sideways, without turning, until B → E. Copy and complete the following:

    A → ..., D → ..., C → ..., AD → ..., BC → ..., DC → ..., parm ABCD → ...

2. Repeat question *1* if tile ABCD is slid upwards so that AD → DH.

3. Use tracing paper to check that ABCD in Figure 12 can be made to coincide with other parallelograms by sliding in various directions, without turning.

4. On squared paper plot the points A (1, 0), B (4, 1), D (3, 3). If ABCD is a parallelogram, write down the coordinates of C. Show on your squared paper that a tiling of parallelograms congruent to ABCD can be extended in all directions.

5. *a* Explain why the quadrilateral with vertices O (0, 0), A (5, 2), B (7, 9), C (2, 7) is a parallelogram.

    b Draw a tiling of parallelograms congruent to OABC.

    c Read off the coordinates of the next three tile corners in the line of

Angles associated with parallel lines

OA. If one of these tile corners is found at the point (x, y), which of the following are true?

(1) $x = \frac{2}{5}y$ (2) $x = \frac{5}{2}y$ (3) $2x = 5y$ (4) $2y = 5x$

6   Find the coordinates of the fourth vertex C of parm ABCD with
- a   A (2, 1), B (5, 2), D (0, 4)   b   A (7, 6), B (10, 7), D (5, 10)
- c   A (80, 72), B (90, 70), D (70, 80)
- d   A (−4, −3), B (4, −1), D (−7, 2)

7   Make a sketch of Figure 12, and shade in the position of the parallelogram tile ABCD after a half turn about:
- a   C   b   the midpoint of BC   c   the midpoint of CD
- d   the midpoint of CF
- e   the point of intersection of the diagonals of parm BEFC.

4 Angles associated with parallel lines

(i) Corresponding angles

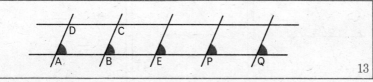

13

Figure 13 shows part of a parallelogram tiling. If we study the congruent parallelograms in the tiling, or think of a parallelogram tile ABCD sliding along the line of AB, we can see that all the coloured angles are equal. Angles alongside parallel lines like this are called *corresponding angles*, since their positions all correspond (in this case *above* the base-line, and *to the right* of the parallels).

For parallel lines corresponding angles are equal.

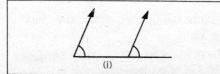
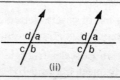

(i)   (ii)   14

Geometry

Figure 14(i) shows the simplest diagram for a pair of corresponding angles, and Figure 14(ii) shows 4 pairs of corresponding angles.

It may help you to recognize corresponding angles if you remember that they always occur in an F-shape, as shown in Figure 15.

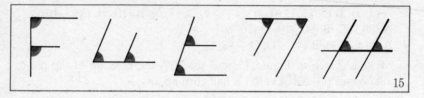

Exercise 6

1. In Figure 16 write down the number of the angle *corresponding* to angle:

   a 1   *b* 2   *c* 3   *d* 4

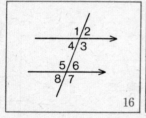
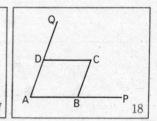

2. In Figure 17, DE is parallel to BC.
   a Name two pairs of corresponding angles.
   b If $\angle BAC = 65°$ and $\angle ADE = 50°$, make a sketch and write in the sizes of all the angles in the figure.

3. In Figure 18, ABCD is a parallelogram.
   a Name two pairs of corresponding angles.
   b If $\angle BAD = 70°$, make a sketch and write in the sizes of all the angles in the figure.

4. Make a list of capital letters which contain parallel lines, and mark in any pairs of corresponding angles.

5. FGH is an isosceles triangle with FG = FH. MN is drawn parallel to GH to meet FG and FH at M and N respectively.

Angles associated with parallel lines

a Name any pairs of corresponding angles in the figure.
b Name another isosceles triangle in the figure.
c If ∠GFH = 40°, make a sketch and write in the sizes of all the angles in the figure.

6 PQRS is a parallelogram with T a point in the side PQ. QU is drawn parallel to TS to meet SR at U. QU is produced to V.

a Name three pairs of corresponding angles in the figure.
b If ∠PTS = 50°, write in the sizes of as many angles as you can.

(ii) *Alternate angles*

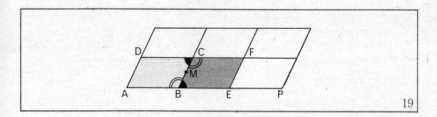

19

If in the parallelogram tiling in Figure 19 we give tile ABCD a half turn about M, the midpoint of BC, then C ↔ B, and parm ABCD ↔ parm FCBE.

In particular, ∠ABC ↔ ∠FCB, and ∠ABC = ∠FCB;
 ∠BCD ↔ ∠CBE, and ∠BCD = ∠CBE.

These pairs of angles are called *alternate angles*, as the angles are *on alternate sides* of the crossing line BC.

For parallel lines alternate angles are equal.

It may help you to recognize alternate angles if you remember that they always occur in a Z-shape, as shown in Figure 20.

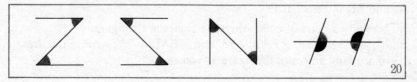

20

Geometry

Exercise 7

1 a In Figure 21, write down the number of the angle alternate to angle:
   (*1*) 4   (*2*) 5

b From the same diagram write down the numbers for:

   (*1*) four pairs of corresponding angles
   (*2*) four pairs of vertically opposite angles
   (*3*) eight pairs of supplementary angles
   (*4*) a set of four equal angles
   (*5*) another set of four equal angles.

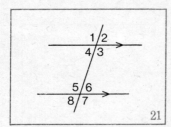
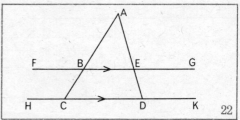

2 a Name as many pairs of alternate angles in Figure 22 as you can.

b If $\angle$DAC = 45° and $\angle$AEB = 70°, make a sketch and write in the sizes of all the angles in the figure.

3 a Sketch a parallelogram ABCD, with its diagonals AC and BD intersecting at O.

b Name four pairs of alternate angles in the figure.

c If $\angle$AOB = 120° and $\angle$OAB = 25°, write in the sizes of as many angles as you can.

4 Make a list of capital letters which contain parallel lines, and mark in any pairs of alternate angles.

5 ABCD is a quadrilateral with AB parallel to DC. The diagonals AC and BD intersect at E.

a Name two pairs of equal alternate angles in the figure.

b If $\angle$AEB = 120°, $\angle$EAB = 25° and $\angle$EAD = 75°, write in the sizes of as many angles in the figure as you can.

Angles associated with parallel lines

Exercise 7B

1 In Figure 23, ABC is an isosceles triangle with AB = AC, and AE is parallel to BC.

   a Name a pair of corresponding angles, and a pair of alternate angles.

   b Why does AE bisect $\angle$CAD?

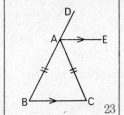 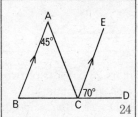 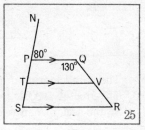

2 In Figure 24, calculate the sizes of angles ABC, ACE and ACB. Give reasons for your answers.

3 Make a sketch of Figure 25, and write in the sizes of all the angles in the diagram.

4 Figure 26(i) shows part of a parallelogram tiling, in which HC and CE are joined. By marking another angle of $a°$, and another of $b°$, prove that $\angle$HCD + $\angle$DCB + $\angle$BCE = 180°. This shows that H, C, E are *collinear* (in the same straight line).

   Figure 26(ii) shows a related tiling of congruent triangles. Copy a diagram like this.

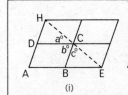 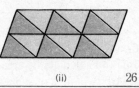 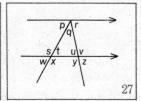

5 a In Figure 27, list the letters of angles which are equal to those marked

     (*1*) *p*   (*2*) *r*   (*3*) *x*   (*4*) *y*.

   b If $p = 56°$ and $q = 44°$, copy the diagram and write in the sizes of all the angles.

5 The area of a parallelogram

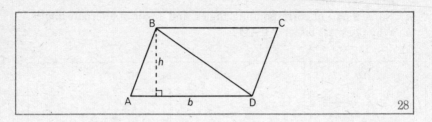

The parallelogram in Figure 28 consists of two congruent triangles.
The area of $\triangle ABD = \frac{1}{2} bh$ square units.
So the area of parm ABCD = $2 \times \frac{1}{2} bh = bh$ square units.

The area of a parallelogram = base × perpendicular height.

On squared paper the area of a parallelogram may be deduced from the area of a surrounding rectangle.

Exercise 8

1. Calculate the areas of the parallelograms in Figure 29. The units are centimetres.

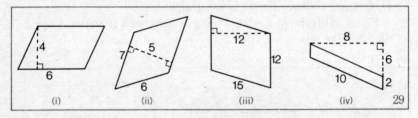

2. Plot the points O (0, 0), A (6, 0), B (8, 3), C (2, 3). Explain why OABC is a parallelogram, and calculate its area

3. *a* Sketch a parallelogram ABCD with AB = 26 mm, AD = 13 mm, and the distance between AD and BC 24 mm.
   b Calculate the area of ABCD, in more than one way if possible.

4. Plot P (1, 1), Q (6, 2), R (7, 6), S (2, 5). What shape is PQRS? Complete a rectangle with diagonal PR and its sides parallel to the *x*- and *y*-axes. Calculate the area of PQRS.

The area of a parallelogram

5   Draw the parallelogram OABC where O is the origin, A is (4, 2), B is (6, 0) and C is (2, −2). Calculate the area of OABC by
  a thinking of the areas of two triangles
  b drawing a rectangle with its sides parallel to the axes.

Geometry

Summary

1   Under a *half turn* about O (or reflection in O):

O is an invariant point;

O is the midpoint of the line joining a point and its image;

a line and its image are parallel, or in the same line, but their directions are opposite.

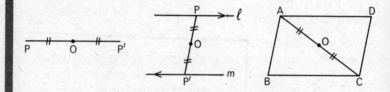

2   A *parallelogram* is formed by a triangle and its image under a half turn about the midpoint of one side. Also:

the opposite sides are equal and parallel;

the opposite angles are equal;

the diagonals bisect each other;

it is conserved under a half turn about the point of intersection of the diagonals;

it can be formed by drawing two pairs of parallel lines.

3   A plane can be covered by a tiling *of congruent parallelograms* containing a *tiling* of congruent triangles.

Summary

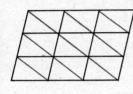 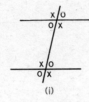 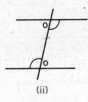

(i) (ii)

4 If a straight line cuts two parallel lines:
   (i) corresponding angles are equal;
   (ii) alternate angles are equal.

5 The area of a parallelogram = base × perpendicular height.

3 Locus, and Equations of a Straight Line

1 The idea of a locus

(i)   (ii)   (iii)

1

Figure 1(i) shows part of a hockey pitch. An attacker must be inside the shaded area (the 'striking circle') before shooting at goal. The shaded area can be regarded as the shooting *locus*. The word 'locus' comes from Latin, and means 'place'.

Figure 1(ii) shows the locus of a goldfish in a bowl, that is the space occupied by the water.

Figure 1(iii) shows the places passed through by a golf ball during its flight, the locus of the ball.

A locus can be thought of as *a set of points*. The path of the golf ball gives a set of points in the form of a curve, the interior of the striking circle a set of points as a two-dimensional region, and the interior of the goldfish bowl a three-dimensional set of points.

Note to the Teacher on Chapter 3

The main aims of this chapter are to introduce the idea of a locus (*Section* 1), to interpret and to use a locus as a set of points (*Sections* 2 and 3) and as the path traced out by a moving object (*Section* 4), and to acquire some familiarity with the equations and graphs of a straight line or a region (*Section* 5).

The traditional definition of *locus* refers to the *path of a moving point*, and the path idea is appropriate when the objective is the description of movement in the real world. But while objects move, points do not, and for mathematical purposes greater clarity is achieved by discussing *locus in terms of sets of points*. The key features are:

(i) a set of points, regarded as a geometrical figure, and called a locus,
(ii) the description of this set in terms of the coordinates of its elements.

In *Section* 2 set-builder notation is introduced because of its usefulness in this context; it is also used in Algebra, Chapter 3, in connection with solution sets of inequations. Exercise 1 deals with circular loci, and Exercise 2 with straight lines. The distinction between the graphical illustrations of the inequations $x \leqslant 3$ and $x < 3$ as shown in Figures 5 and 6 is worth emphasizing, as it will occur in various parts of the course. To a linear equation corresponds a straight line, and to a linear inequation corresponds a plane region, a 'half-plane'. Exercise 3B introduces particular cases of the hyperbola and ellipse, as well as the classical perpendicular bisector locus.

Section 3 relates the pupils' knowledge of the intersection of sets to the intersection of the corresponding loci, and again the emphasis is on straight lines and regions of the plane.

In *Section* 4, the examples given in Exercise 6 might well be supplemented by suggestions from the pupils.

The equations of a straight line will be studied again in more detail and more formally later in the course, but pupils will have to be familiar with some aspects of these for the work on *Systems of Equations* in the algebra of Book 4, and it falls into place naturally in *Section* 5 of the present chapter. The approach is made through 'tile corners' as in earlier chapters, and should lead the pupils easily enough at this stage to accept the statement concerning straight lines and their equations. The development is made in four stages;

(i) $y = x$ and $y = -x$, with an introduction of the regions $y > x$ and $y < x$ in questions *9* and *10* of Exercise 6,
(ii) $y = x+c$ (iii) $y = mx$ (iv) $y = mx+c$.

Formal proofs concerning these equations are given at a later stage. The method suggested at the end of Exercise 9 for sketching straight lines with given equations will often be found to be useful.

Many of the sketches in *Section* 5 can be made on small areas of squared paper; 5-mm squares will be best in general.

(facing page 115)

2 Sets of points

In Figure 2(i) the coloured region A consists of the set of points which are 1 cm or less from the origin.

We can describe this in *set-builder notation*, $A = \{P: OP \leq 1 \text{ cm}\}$, to be read as 'the set of points P such that OP is less than or equal to 1 centimetre'.

This set of points is called the *locus of points* in the plane which are 1 cm or less from O.

In Figure 2(ii) the coloured region B consists of the set of points which are 1 cm or more from O, i.e. $B = \{P: OP \geq 1 \text{ cm}\}$. This region is the locus of points which are 1 cm or more from O.

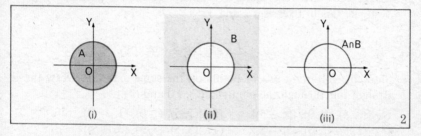

$A \cap B = \{P: OP = 1 \text{ cm}\}$. This is the set of points 1 cm from O. These points form the circumference of the circle where the two regions A and B meet, as shown in Figure 2(iii).

Exercise 1

1 a Mark a point C near the middle of a page in your notebook, and then draw a circle with centre C and radius 2 cm.
   b Mark in red at least ten points 2 cm or less from C.
   c Mark in blue at least ten points 2 cm or more from C.
   d Where are points that can be marked both in red and blue?
   e How many points could be marked in *b*, *c* and *d* above?
   f Shade the red region. Call it R. Complete $R = \{P: CP \leq ...\}$
   g Shade the blue region. Call it B. Complete $B = \{P: CP ...\}$
   h The set of points in both the red and blue regions $= R \cap B = \{P: ...\}$

Geometry

2 Draw separate diagrams to show the locus of points:
- *a* 3 cm from O
- *b* 3 cm or more from O
- *c* 4 cm from O
- *d* 4 cm or less from O

3 Draw separate diagrams to show the locus of P defined by:
- *a* {P: OP = 5 cm}
- *b* {P: OP ⩾ 5 cm}
- *c* {P: OP = 2·5 cm}
- *d* {P: OP ⩽ 2·5 cm}

4 Show *in the same diagram* the locus of points defined by
- *a* {P: OP = 1 cm}   *b* {P: OP = 2 cm}   *c* {P: 1 ⩽ OP ⩽ 2 cm}

5 Show by shading the locus of points given by {P: 5 cm ⩽ OP ⩽ 7 cm}

6 Describe the locus *in three dimensions* of all points P which are:
- *a* 6 cm from a fixed point O, i.e. {P: OP = 6 cm}
- *b* less than 6 cm from O
- *c* more than 6 cm from O.

Coordinates are very useful for describing sets of points; we investigate this in Exercises 2 and 3.

Exercise 2

1 *a* Show the following sets of points on the same diagram. Draw the straight line through the points in (2), (3) and (4).
   (*1*) {(0, 5), (0, 0), (0, −5)}   (*2*) {(1, 5), (1, 0), (1, −5)}
   (*3*) {(2, 5), (2, 0), (2, −5)}   (*4*) {(3, 5), (3, 0), (3, −5)}

   b Describe the locus of all points with x-coordinate:
   (*1*) 0   (*2*) 1   (*3*) 2   (*4*) 3   (*5*) positive.

2 *a* Repeat question *1a* on another diagram for the following sets of points:
   (*1*) {(0, 5), (0, 0), (0, −5)}   (*2*) {(−1, 5), (−1, 0), (−1, −5)}
   (*3*) {(−2, 5), (−2, 0), (−2, −5)}
   (*4*) {(−3, 5), (−3, 0), (−3, −5)}

   b Describe the locus of all points with x- coordinate:
   (*1*) −1   (*2*) −2   (*3*) −3   (*4*) negative.

3 *a* Study the following text, and Figures 3, 4, 5 and 6.
   b Copy the Figures into your notebook, and describe them in words or using set notation.

   Figure 3(i) shows in colour the locus of points whose x-coordinates are positive or zero.

Sets of points

Figure 3(ii) shows in colour the locus of points whose x-coordinates are negative or zero.

Figure 3(iii) shows in colour the locus of points whose x-coordinates are zero.

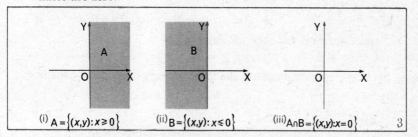

(i) $A = \{(x,y): x \geq 0\}$ (ii) $B = \{(x,y): x \leq 0\}$ (iii) $A \cap B = \{(x,y): x = 0\}$

3

The set of points with x-coordinate zero, i.e. $\{(x, y): x = 0\}$, forms the y-axis. The equation of this line is $x = 0$.

Figure 4 shows in colour the set of points defined by $\{(x, y): x = 3\}$, i.e. the straight line with equation $x = 3$.

Figure 5 shows in colour the region $\{(x, y): 0 < x < 3\}$. The boundary lines are broken (or dotted) to show that they do not belong to the set.

Figure 6 shows in colour the region $\{(x, y): 0 \leq x \leq 3\}$; here the boundary lines (shown as solid lines) *do* belong to the set.

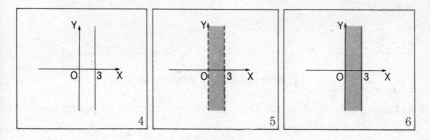

Exercise 3

1 Show the following sets of points on the same diagram:
   a $\{(x, y): x = 1\}$ b $\{(x, y): x = 4\}$
   c $\{(x, y): y = 5\}$ d $\{(x, y): y = -5\}$

2 Show the following sets of points on the same diagram:
   a $\{(x, y): x = 0\}$ b $\{(x, y): y = 0\}$
   What special lines are these?

Geometry

3 Sketch on the same diagram the lines with equations:

    a $x = 3$ *b* $x = 6$ *c* $y = 7$ *d* $y = -4$

4 Sketch the following loci (plural of locus) on the same diagram, and say what you notice about them.

    a $\{(x, y): x = 5\}$ *b* $\{(x, y): x = 0\}$ *c* $\{(x, y): x = -3\}$

5 Repeat question **4** for the following loci:

    $\{(x, y): y = 3\}$ $\{(x, y): y = 0\}$ $\{(x, y): y = -3\}$

6 Show on separate diagrams the following sets of points, by shading:

    a $\{(x, y): x \geq 0\}$ *b* $\{(x, y): y \leq 0\}$
    c $\{(x, y): x > 5\}$ *d* $\{(x, y): y < 3\}$

7 Show, by shading, the following loci, on separate diagrams:

    a $\{(x, y): 0 < y < 2\}$ *b* $\{(x, y): 0 \leq y \leq 2\}$
    c $\{(x, y): -1 \leq x < 1\}$ *d* $\{(x, y): -2 < x < 6\}$

8 Use set notation to define the locus of points in the coloured regions of Figure 7.

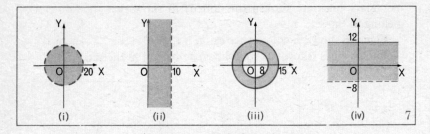

Exercise 3B. Various loci

1 *a* Copy and complete this table giving the distances of points from axes OX and OY.

Distance from OX	1	1·5	2	3		6	8	12
Distance from OY	12	8	6		3			
Product of distances	12	12	12	12	12	12	12	12

    b On 5-mm squared paper plot the points (1, 12), (1·5, 8), (2, 6), etc.

    c Draw a smooth curve through these points to show part of the locus of all such points (a *hyperbola*).

2 *a* Draw a circle with diameter AB 10 cm long. Mark P on the circle and draw PN perpendicular to AB. Mark Q, the midpoint of PN.

Intersection of sets

    b Take at least ten other positions of P on the circle; draw PN for each, and mark the various positions of Q.

    c Draw a smooth curve through these positions of Q to show the locus of all such points (an *ellipse*).

3 Draw a line AB 6 cm long. Using compasses, take centre A and radius more than half AB, and draw arcs above and below AB. With centre B and the same radius, draw arcs to cut the first two at P and Q. Repeat for a different radius to get points R and S. Repeat for another radius to get T and U. Sketch, and describe, the locus of all points which are equidistant from A and B.

3 Intersection of sets

Introductory Exercise

1 *a* There is a buried treasure in a rectangular field ABCD. One man knows that the treasure is 20 metres from A. He marks in a diagram some possible positions of the treasure. Describe the set of all possible positions, and mark it in a scale diagram.

    b Another man B knows that the treasure is 15 metres from the side AB of the field. Describe the set of all possible positions as he knows them, and mark it in the same diagram.

    c What is the intersection of the two sets?

2 The centre circle of a football pitch has radius 10 metres. Show on a diagram the set of points which are 10 metres from the centre of the field, and which are 5 metres from the centre line. Describe this set, using the symbol for intersection of sets.

3 In a diagram like that for question 2 show $P \cap Q$, where P denotes the set of points not more than 10 metres from the centre, and Q the set of points 5 metres from the centre line.

4 ABCD is a rectangle in which AB is 6 cm long and BC is 4 cm long. Indicate on separate sketches the following sets of points *inside* the rectangle.
P is the set of points which are less than 1 cm from AB.
Q is the set of points which are 1 cm from AB.
R is the set of points which are more than 1 cm from AB.
S is the set of points which are less than 1 cm from AD.

Geometry

T is the set of points which are 1 cm from AD.
U is the set of points which are more than 1 cm from AD.
Describe the following intersections in words:

a $Q \cap T$ *b* $P \cap S$ *c* $Q \cap S$ *d* $R \cap U$

In Figure 8, L is the set of points with x-coordinate 3, and M is the set of points with y-coordinate 1. L and M intersect in the set containing only the point (3, 1).

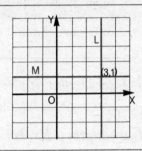

8

$L = \{(x, y): x = 3\}, \quad M = \{(x, y): y = 1\}.$
$L \cap M = \{(x, y): x = 3 \text{ and } y = 1\} = \{(3, 1)\}.$

Exercise 4A

1 *a* Sketch the sets of points $L = \{(x, y): x = 5\}$ and $M = \{(x, y): y = 5\}$ on squared paper.
    b What set is given by $L \cap M$?

2 Repeat question *1* for $L = \{(x, y): x = -2\}$ and $M = \{(x, y): y = -1\}$.

3 Repeat question *1* for $L = \{(x, y): x = 0\}$ and $M = \{(x, y): y = 2\}$.

4 *a* On the same diagram sketch the lines with equations:
$$x = 6, \quad y = 6, \quad x = 2 \quad \text{and} \quad y = -1.$$
    b Give the points of intersection of pairs of the lines.
    c Which pairs do not intersect on your diagram?

5 *a* Show the following sets of points on the same diagram:
$P = \{(x, y): x = 3\}, \quad\quad Q = \{(x, y): y = -2\},$
$R = \{(x, y): x = -4\}, \quad\quad S = \{(x, y): y = 5\}$
    b Use set notation to give $P \cap Q$, $Q \cap R$, $R \cap S$ and $S \cap P$.
    c What is $P \cap R$, and $Q \cap S$?

Intersection of sets

6 *a* Show, by shading, on the same diagram, the regions:
$A = \{(x, y): 1 \leqslant x \leqslant 8\}$ and $B = \{(x, y): -1 \leqslant y \leqslant 6\}$
b Colour the region $A \cap B$. Calculate its area.

7 Repeat question 6 for the regions $C = \{(x, y): -2 < x < 2\}$ and $D = \{(x, y): -2 < y < 2\}$

Exercise 4B

1 *a* Sketch the sets of points $L = \{(x, y): x = 3\}$ and $M = \{(x, y): y = -1\}$ on squared paper.
b What set is given by $L \cap M$?

2 Repeat question *1* for $L = \{(x, y): x = -5\}$ and $M = \{(x, y): y = 0\}$

3 *a* On the same diagram sketch the lines with equations
$$x = 4, \quad y = -2, \quad y = x.$$
b Give the points of intersection of pairs of the lines.

4 *a* Show the following sets of points on the same diagram
$A = \{(x, y): x = 6\} \quad B = \{(x, y): x = 2\},$
$C = \{(x, y): y = 3\}, \quad D = \{(x, y): y = -1\}$
b Use set notation to give $A \cap B, A \cap C, A \cap D, B \cap C, B \cap D$ and $C \cap D$.

5 *a* Show by shading the locus of points given by:
$P = \{(x, y): 1 < x < 3\}$ and $Q = \{(x, y): 2 < y < 6\}$
b Colour the region $P \cap Q$. Calculate its area.

6 Repeat question 5 for $P = \{(x, y): -5 < x < 0\}$ and $Q = \{(x, y): -2 < y < 2\}$

7 $A = \{(x, y): y < 2\}, \quad B = \{(x, y): y = 2\}, \quad C = \{(x, y): y > 2\},$
$D = \{(x, y): x < 2\}, \quad E = \{(x, y): x = 2\}, \quad F = \{(x, y): x > 2\}$
Show by separate sketches the following intersections:
a $B \cap E$ *b* $B \cap F$ *c* $E \cap C$ *d* $A \cap D$ *e* $A \cap F$

8 $P = \{(x, y): y \geqslant x\}, \quad Q = \{(x, y): y = x\}, \quad R = \{(x, y): y \leqslant x\}.$
a Show the loci given by P, Q and R in separate sketches.
b Comment on $P \cap Q, Q \cap R$ and $P \cap R$.
c $S = \{(x, y): y > 1\}, \quad T = \{(x, y): y = 1\}, \quad U = \{(x, y): y < 1\}.$
Sketch:
(1) $Q \cap T$ (2) $Q \cap S$ (3) $R \cap S$ (4) $P \cap T$ (5) $P \cap U$.

Geometry

4 Paths of moving objects

The path of a moving object gives a set of points, and the word 'locus' is often used to describe this path.

For example, if the wheel in Figure 9 is turned about its centre O, the locus of the valve V is the circle shown in colour.

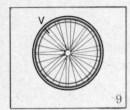
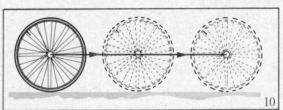

If the wheel is rolled along level ground the locus of the centre is the horizontal line shown in colour in Figure 10.

Exercise 5

Sketch, or describe, or indicate in some way, the shapes of the paths followed by the moving objects in the following questions.

1   A boy at the fairground
  a   on a roundabout   *b*   on the big wheel
  c   on a helter-skelter.

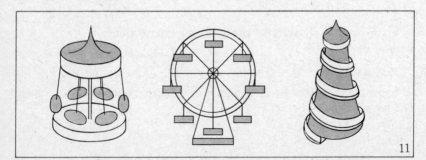

Paths of moving objects

2   A lift or elevator in a tall building.
3   A golf ball:
  a on a level putting green   *b* after being driven from a tee.
4   A speck of dust on a record-player disc.
5   A record-player needle.
6   A child on a swing.
7   The tip of the hand of a clock.
8   A model aircraft flying at the end of an anchored wire.
9   The tip of an aircraft propeller
  a when the aircraft is stationary   *b* when the aircraft is in flight.
10 The earth in the course of one year.

Exercise 5B

1   An aircraft flies due east from take-off, is refuelled in the air, and continues to fly until it returns to base. Describe its path.

2   Repeat question *1* for an aircraft which sets off due north and flies at a constant height. What points on earth will it fly over?

3   In the playground a boy walks 10 m east then 10 m north, then 10 m west, then 10 m south. Sketch his path. What shape is it?

4   Sketch the locus of the centre of a wheel of radius 5 cm which:
  a rolls along a horizontal rail
  b rolls round another wheel of radius 5 cm
  c rolls round the inside of a rectangular box whose base measures 15 cm by 12 cm   (The wheel lies flat on the base, and rolls against the side walls.)
  d rolls round the outside of the box in *c*.

5 *a* A weight is attached to the end A of a string OA 100 cm long, hanging vertically from O. The weight is pulled to one side, and is then allowed to swing to and fro. Sketch and describe the path of the weight.
  b If an obstacle is placed in the path of the string at B, a point 50 cm below O, sketch the path of the weight now.

Geometry

5 Equations of straight lines

(i) $y = x$ and $y = -x$

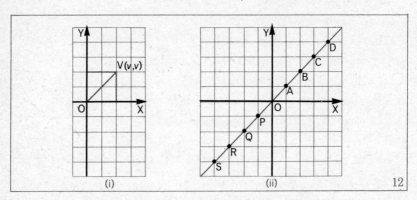

In Figure 12(i), O and V are opposite vertices of a square, and OV makes an angle of 45° with the x-axis.

V is the point (v, v), i.e. $V \in \{(x, y): y = x\}$

In Figure 12(ii), A is $(1, 1)$, B $(2, 2)$, C $(3, 3)$, D $(4, 4)$. Each of these points is the vertex of a square like V in Figure 12(i), so that OA, OB, OC and OD each make an angle of 45° with OX; it follows that A, B, C and D all lie on the straight line through O at 45° to OX. The same is true for P, Q, R and S, and in fact for all such points, so this line is the locus of all these points.

The set $\{(x, y): y = x\}$, i.e. the set of all points for which $y = x$, consists of a straight line through the origin, at 45° to OX, as shown in Figure 12, and is often referred to as the line with equation $y = x$.

Exercise 6

1 a Plot the points $(10, 10)$, $(5, 5)$, $(-5, -5)$ and $(-10, -10)$ on squared paper.

b Draw the straight line through these points, and write down its equation.

2 Which of the following points lie on the line in question *1b*?
A $(2, 2)$, B $(0, 2)$, C $(-1, -1)$, D $(80, 80)$, E $(-100, -100)$, O $(0, 0)$

Equations of straight lines

3. Given that the following points lie on the line with equation $y = x$, write down the values of $a, b, c, \ldots$
P (6, a), Q (−7, b), R (1000, c), S (−15, d), T (e, 0), U (f, 0·1)

4. a. Plot the points A (1, −1), B (2, −2), C (3, −3), D (4, −4).
   b. Draw in squares as in Figure 12. Why do A, B, C, D all lie on the same straight line? Draw this line.
   c. Which of the following correctly define the set of points on the line you have drawn?

   (1) $\{(x, y): y = x\}$ (2) $\{(x, y): y = -x\}$
   (3) $\{(x, y): x - y = 0\}$ (4) $\{(x, y): x + y = 0\}$
   (5) $\{(x, y): x = -y\}$ (6) $\{(x, y): x = y\}$

5. Which of the following points lie on the line in question **4b**?
P (5, −5), Q (10, 10), R (−5, −5), S (−5, 5), T (100, −100), O (0, 0)

6. Given that each of the following points lies on the line with equation $y = -x$, write down the values of $a, b, c, \ldots$
A (10, a), B (−2, b), C (1, c), D (d, 1000), E (e, 0), F (f, −25)

7. a. Draw the line with equation $y = x$.
   b. Plot the points A (4, 2), B (5, 1), C (3, 3), D (3, 5), E (1, 2), F (−4, −1), G (−2, −5).
   c. Which of these points lie *above* the line, and which lie *below* the line?
   d. If the point (p, q) lies above the line, which of the following is true?
   (1) $p > q$ (2) $p = q$ (3) $p < q$

8. a. Draw the line with equation $y = -x$.
   b. If (p, q) lies below the line, which of the following is true?
   (1) $p < -q$ (2) $q < -p$ (3) $q < p$ (4) $p < q$

9. Figure 13 shows the three sets of points
$$\{(x, y): y > x\}, \quad \{(x, y): y = x\}, \quad \{(x, y): y < x\}.$$
Copy the diagram, and write down the coordinates of two points in each set.

Geometry

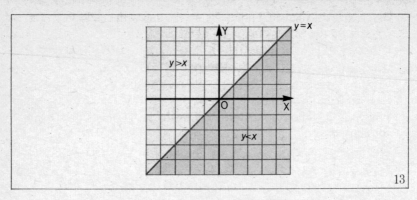

10 Draw a diagram similar to Figure 13 to show the three sets
$\{(x, y): y > -x\}$, $\{(x, y): y = -x\}$, $\{(x, y): y < -x\}$.

*   *   *

(ii) $y = x + c$

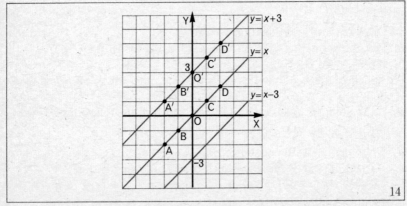

In Figure 14, the line $y = x$ has been drawn, and five points have been marked on this line:

A$(-2, -2)$, B$(-1, -1)$, O$(0, 0)$, C$(1, 1)$ and D$(2, 2)$.

From these points five more points have been obtained by increasing the y-coordinates by 3:

A'$(-2, 1)$, B'$(-1, 2)$, O'$(0, 3)$, C'$(1, 4)$ and D'$(2, 5)$.

These points lie on a line parallel to $y = x$, and since their coordinates satisfy the equation $y = x + 3$, this is the equation of the

Equations of straight lines

line on which they lie. Notice that this line cuts the *y*-axis 3 units above O, at (0, 3).

The line with equation $y = x-3$ is also shown in the diagram; notice that this line cuts the *y*-axis 3 units below O, at (0, −3).

All points (x, y) for which $y = x+c$ (where c is constant) lie on a straight line parallel to $y = x$, through the point $(0, c)$.

Exercise 7

1. Draw the line with equation $y = x$. On the same diagram draw the lines with equations $y = x+1$ and $y = x-1$; give the coordinates of the points on the *y*-axis through which these lines pass.

2. Draw the line through A (0, 5) parallel to the line $y = x$. What is the equation of this line?

3. Repeat question 2 for the line through B (0, −6) parallel to the line $y = x$.

4. Say as much as you can about the lines with equations:
   a $y = x$ b $y = x+100$ c $y = x-10$

5. a Sketch the lines with equations $y = x$ and $x = 4$.
   b Write down the coordinates of the point of intersection of these lines.

6. Repeat question 5 for the equations $y = x+2$ and $x = 3$.

7. Repeat question 5 for the equations $y = x-3$ and $x = 5$.

8. Repeat question 5 for the equations $y = x+4$ and $y = 5$.

Geometry

(iii) $y = mx$

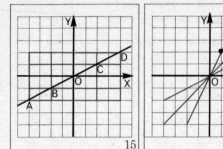
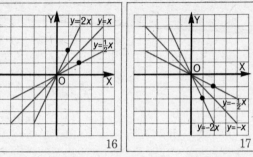

15 16 17

In Figure 15, A is the point $(-4, -2)$, B $(-2, -1)$, C $(2, 1)$, D $(4, 2)$. These points are at corners of the rectangular tiling shown, and all lie on the same straight line through O. For every point on this line, $y = \tfrac{1}{2}x$, so the locus of all such points is given by $\{(x, y): y = \tfrac{1}{2}x\}$, the straight line with equation $y = \tfrac{1}{2}x$.

Figures 16 and 17 show the lines with equations:
$$y = 2x, \quad y = \tfrac{1}{2}x, \quad y = -\tfrac{1}{2}x \text{ and } y = -2x.$$

Notice the coordinates of the tile corners marked in the diagrams, $(1, 2)$, $(2, 1)$, $(2, -1)$, $(1, -2)$, and the coefficients of x in the 2, $\tfrac{1}{2}$, $-\tfrac{1}{2}$, -2, equations:

These coefficients, given by the ratio

$$\frac{y\text{-coordinate of a point on line through O}}{x\text{-coordinate of the same point}}$$

provide a measure of the *slope*, or *gradient*, of the line.

All points (x, y) for which $y = mx$ (where m is constant) lie on a straight line through the origin, with gradient m.

Exercise 8

1 *a* Plot the points $(1, 2), (2, 4), (3, 6), (4, 8)$.
  b Draw the straight line through these points, and write down its equation.
2 Repeat question *1* for the points $(-4, -2), (-2, -1), (2, 1), (4, 2)$.
3 Repeat question *1* for the points $(3, 2), (6, 4), (9, 6)$.
4 Repeat question *1* for the points $(2, -1), (4, -2), (6, -3)$.
5 Sketch the lines with the following equations, on the same diagram:
  a $y = x$ *b* $y = 2x$ *c* $y = 3x$ *d* $y = \tfrac{1}{3}x$
  Write down the gradient of each line.

Equations of straight lines

6 Write down the equations of the lines through the origin, with gradients:

 a 1   *b* 5   *c* −1   *d* $\frac{1}{10}$ *e* −10   *f* 123

7 Sketch the line through the origin and the given point in each of the following; then write down the equation of the line.

 a A (2, 2)   *b* B (3, 6)   *c* C (8, 4)   *d* D (5, 2)   *e* E (1, 6)

8 Repeat question 7 for the points:

 a P (4, −4)   *b* Q (4, −2)   *c* R (−5, −5)
 d S (−5, −1)   *e* T (8, 0)

(iv) $y = mx + c$

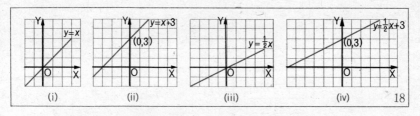

Figure 18(iv) combines the ideas illustrated in (i), (ii) and (iii) and shows the graph of the line with equation $y = \frac{1}{2}x + 3$.

This line cuts the *y*-axis at the point (0, 3), and has gradient $\frac{1}{2}$.

All points (x, y) for which $y = mx + c$ lie on a straight line through the point $(0, c)$, with gradient m.

Exercise 9

1 Sketch the straight lines with equations:

 a $y = x$ *b* $y = x+1$ *c* $y = 2x+1$

2 Repeat question *1* for the equations:

 a $y = 2x$ *b* $y = 2x+3$ *c* $y = 2x-3$

3 Repeat question *1* for the equations:

 a $y = \frac{1}{2}x$ *b* $y = \frac{1}{2}x+4$ *c* $y = \frac{1}{2}x-4$

4 Repeat question *1* for the equations:

 a $y = x+2$ *b* $y = 3x+1$ *c* $y = 2x-5$

5 Write down the gradients of the lines with the following equations, and also the coordinates of the points where they cut the *y*-axis:

 a $y = -x$ *b* $y = -x+3$ *c* $y = -2x-4$ *d* $y = -3x+1$

6 Sketch the lines with the equations in question 5.

Geometry

A useful way to sketch a straight line quickly is to plot the two points where it cuts the x- and y-axes, and to draw the line through these points.

Example. Sketch the line with equation $y = 3x-6$, and shade the region showing the locus of points given by $\{(x, y): y \geqslant 3x-6\}$.
If $x = 0$, $y = -6$.
If $y = 0$, $3x-6 = 0$, so $x = 2$.
The line cuts the axes at $(0, -6)$ and $(2, 0)$.
The line and the required region are shown in Figure 19.

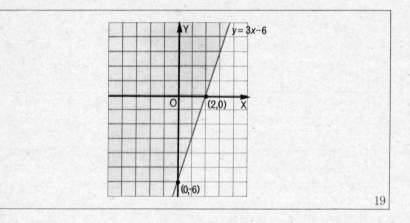

Exercise 10

1 Sketch the lines with equations:
   a $y = x+6$ *b* $y = 2x-4$ *c* $2y = 3x-6$

2 Repeat question *1* for the equations:
   a $x+y = 3$ *b* $2x+5y = 10$ *c* $3y-x = 3$

3 On separate diagrams shade the regions given by:
$$A = \{(x, y): y > x\}, \quad B = \{(x, y): y > x+2\},$$
$$C = \{(x, y): y < -x\}, \quad D = \{(x, y): y < x-3\}$$

4 Combine your answers to question *3* to illustrate or describe:
   a $A \cap B$ *b* $A \cap C$ *c* $A \cap D$ *d* $C \cap D$

Summary

1 A **locus** can be regarded as a set of points defined in some way, an important special case being the path traced out by a moving object.

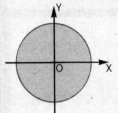 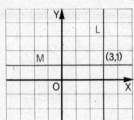 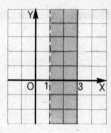

{P: OP ⩽ 1 cm} $L = \{(x, y): x = 3\}$, $\{(x, y): 1 < x \leqslant 3\}$
                       $M = \{(x, y): y = 1\}$
                       $L \cap M = \{(3, 1)\}$

2 Equations of a straight line

(i) $y = x$ is the equation of a straight line through the origin, at 45° to OX.

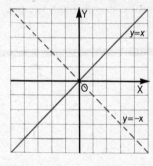 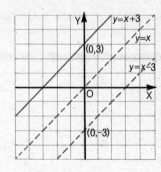

(ii) $y = x + c$ is the equation of a line parallel to $y = x$, through the point $(0, c)$.

Geometry

(iii) $y = mx$ is the equation of a line through the origin, with gradient m.

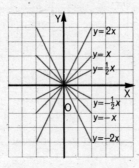 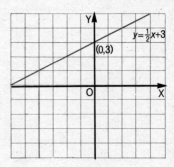

(iv) $y = mx + c$ is the equation of a line through the point $(0, c)$, with gradient m, parallel to the line $y = mx$.

Revision Exercises

Revision Exercises on Chapter 1
Reflection

Revision Exercise 1A

1. Figure 1 shows a square with two of its axes of symmetry, and an equilateral triangle. Make three copies of the diagram, and draw in another triangle in each one so that the figure will have
   a. bilateral symmetry about (*1*)PQ (*2*)RS
   b. half turn symmetry about O.

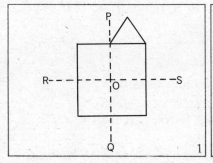
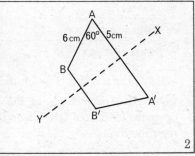

2. In Figure 2, A'B' is the image of AB in XY. Write down
   a. the lengths of A'B' and AA'
   b. the sizes of angle AA'B' and the angle between AA' and XY.

3. If AB is produced in Figure 2 to meet XY at T, calculate the size of ∠ATA'.

4. In Figure 3, AX bisects ∠BAC. Copy the diagram, and draw the image P'Q' of PQ in XA. Join PP' and QQ', and name four isosceles triangles in your diagram.

Geometry

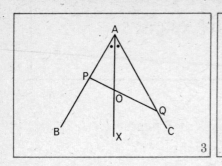
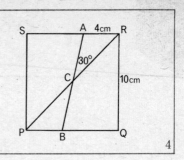

5 In Figure 4, PQRS is a square of side 10 cm. RA = 4 cm, and ∠ACR = 30°. Show the images A' and B' of A and B under reflection in PR. Write down the length of A'Q, and the size of ∠AA'B'.

6 Explain why O (0, 0), A (5, 5), B (10, 0), C (5, −5) form a square. Give the coordinates of the vertices of the image squares under reflection in: *a* the *x*-axis *b* the *y*-axis.

7 Copy and complete this table for reflections in the lines $x = 4$ and $x = 10$:

Given point A	(3, 6)	(0, 2)	(1, −3)	(4, 0)	(7, 1)	(12, 3)
A', image of A in $x = 4$						
A'', image of A' in $x = 10$						

What is the relation between the coordinates of A and A''?

8 PQRS is a rhombus with PQ = 8 cm and ∠QRS = 70°. Write down the lengths of QR, RS and PS, and the sizes of all the angles in the rhombus.

9 Draw a tiling of four rhombuses which make one larger rhombus. Draw the axes of symmetry, and shade a rectangle in the tiling, explaining why it is a rectangle.

10 ABCD is a kite with ∠BAD = 78° and ∠BCD = 40°. Calculate the sizes of angles ABC and ADC.

11 Construct a kite with sides 4 cm and 8 cm long, and shorter diagonal 6 cm long.

12 Describe the shape formed by joining P (4, 6), Q (6, 16), R (8, 6), S (6, 0), and calculate its area.

Revision Exercises on Chapter 1 135

13 P, Q, R of question *12* remain fixed, but S moves to T so that PQRT is a rhombus. Write down the coordinates of T, and the area of the rhombus.

14 Show that A (4, 3), B (8, 5), C (12, 3), D (8, 1) form a rhombus when joined, and give the coordinates of the images of the vertices of the rhombus under reflection in: *a* the *x*-axis   *b* the *y*-axis.

15 Construct an isosceles triangle ABC with base BC 6 cm long and base angles of size 70°. Construct the altitude from A, and the bisectors of the angles at B and C. What do you find?

Revision Exercise 1B

1 State which of the shapes in Figure 5 have bilateral symmetry and give the number of axes of symmetry, and which have half turn symmetry. (You may wish to use tracing paper.)

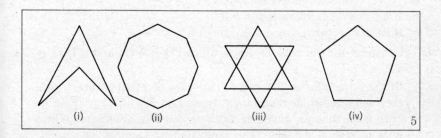

2 Triangle ABC is right-angled at A. Sketch and name the kind of combined figures obtained from △ABC and its image under reflection in
a the line of AB   *b* the line of AC   *c* the line of BC
d the line of AB followed by reflection in the line of AC.

3 Find the images of $(-5, 8)$, $(4, 4)$, $(10, 0)$ and (a, b) under reflection in the *y*-axis followed by reflection in the line $x = 5$.

4 Find the images of the given points in question *3* under reflection in the *x*-axis followed by reflection in the *y*-axis.

5 *a* What line do the points $(0, 0)$, $(1, 1)$, $(4, 4)$, $(8, 8)$ lie on?
  b Write down the coordinates of the images of O $(0, 0)$, B $(4, 0)$, C $(4, 2)$, D $(0, 8)$, E $(6, 8)$, F (p, q) under reflection in the line in *a*.

6 LMNP is a rhombus. It is folded so that N → P and M → M′,

Geometry

where M' lies on LM. Show this in a sketch. Name angles equal to PNM and NML.

7 The area of a rhombus is 30 cm². If one of its diagonals is 6 cm long, calculate the length of the other diagonal.

8 Explain why A (0, 2), B (5, 0), C (10, 2), D (5, 4) form a rhombus when joined. Draw a tiling of rhombuses based on ABCD, and write down the coordinates of four tile corners in the line of AD. If R (x, y) is one such corner, which of these is true?

   a $y = \frac{2}{5}x$ b $y = \frac{5}{2}x$ c $y = \frac{2}{5}x + 2$ d $y = \frac{5}{2}x + 2$

9 A kite has diagonals of lengths x cm and y cm, and the area of the kite lies between $12\frac{1}{2}$ cm² and 18 cm². Show that $25 < xy < 36$.

10 If in question 9, $x = y$, and the diagonals bisect each other, what shape is the figure? If its area is 32 cm², what is the value of x?

11 S = {squares}, B = {rhombuses}, R = {rectangles}.
  a If $a \in R$, does it follow that $a \in S$?
  b If $b \in S$, does it follow that $b \in B$?
  c If $c \in B \cap R$, what can you say about c?

12 Construct the rhombus PQRS in which PQ = 4 cm and $\angle$PQS = 40°.

13 'If the perpendicular bisectors of two sides of a triangle are drawn, they will meet inside the triangle, or on one side of the triangle, or outside the triangle.' Investigate this statement by constructing these lines in several different shapes of triangles (acute-angled, right-angled, obtuse-angled).

Revision Exercises on Chapter 2
The Parallelogram

Revision Exercise 2A

1 Draw the letter **A**, and in the same diagram draw its image under a half turn about its highest point.

2 A line AB 10 cm long is mapped by a half turn about a point O to a line CD. Explain, with the aid of a sketch, where the position of O is with respect to AB and CD.

Revision Exercises on Chapter 2

P is a point on AB 6 cm from B. Where is the image of P?
Q is a point on DC 2 cm from C. Where is the image of Q?

3 Under a half turn about the origin, what are the images of the points $(3, 5)$, $(-2, 4)$, $(1, -5)$, (a, b) and $(-p, q)$?

4 Under a half turn about the point $(3, 2)$, what are the images of $(5, 6)$, $(2, 4)$, $(-1, 3)$ and $(-2, -1)$?

5 Write down any capital letters of the alphabet which have a centre of symmetry. Mark the centre clearly in each case.

6 State clearly the effect of reflecting a straight line AB in a point O which: *a* lies on AB   *b* does not lie on AB.

7 O is the centre of a rectangle ABCD in which AB = 10 cm and BC = 6 cm. A line POQ meets AB at P and DC at Q. If PB = 3 cm find: *a* the length of QC   *b* the area of quadrilateral PBCQ.

8 State whether each of the following statements is true or false for every parallelogram.
  a The opposite sides are equal.
  b Each diagonal divides the parallelogram into two congruent triangles.
  c The diagonals bisect the angles through which they pass.
  d The sum of its angles is 360°.
  e The diagonals are perpendicular to each other.
  f If two diagonals are drawn, the parallelogram is divided into four congruent triangles.
  g A parallelogram has no axis of symmetry.

9 A quadrilateral ABCD has two pairs of opposite sides parallel. How would you find its centre of symmetry? In how many ways does it fit its outline?

10 Draw a tiling of parallelograms, and investigate the truth of the following statements. (Use tracing paper if it helps you.)
  a Every parallelogram in the tiling can be moved into the space of any other parallelogram by sliding it without turning.
  b Every parallelogram in the tiling can be turned into the space of any parallelogram (including itself) by choosing a suitable centre for a half turn.

11 Sketch a pair of parallel lines and a line crossing them. If one of the acute angles in the figure is 47°, find the size of every other angle in the figure.

Geometry

12 In Figure 6 the lines marked with the arrows are parallel. Write down as many equations as you can involving any of x, y, z and u.

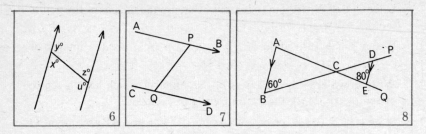

13 In Figure 7, AB is parallel to CD. Name as many pairs of:
   a equal angles   *b* supplementary angles as you can.

14 In Figure 8, AB ∥ DE. ∠ABC = 60° and ∠DEC = 80°. Calculate the sizes of all the other angles in the figure.

15 Sketch a rectangle ABCD with the diagonal AC drawn. If ∠BAC = 34°, what is the size of each angle in the figure?

16 Calculate the area of the parallelogram with vertices (0, 0), (0, 5), (4, 7) and (4, 2).

Revision Exercise 2B

1 State which letter is formed by each of the shapes in Figure 9 and its image under reflection in the point O.

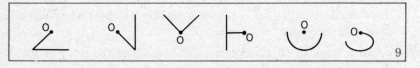

2 Draw a letter **F** by plotting the points (1, 1), (1, 2), (1, 3), (2, 2) and (3, 3) on a coordinate diagram. Draw the images of the letter under reflection in the following points:
   a (0, 0)   *b* (3, 0)   *c* (−1, 2)

3 What are the images of the following points under a half turn about (−3, −2)?   *a* (6, 4)   *b* (−3, 2)   *c* (5, −2)   *d* (a, b)

4 A is the point (−2, 3) and B is (4, −5). Find the images of (5, −3) and (p, q) under a half turn about the midpoint of AB.

Revision Exercises on Chapter 2

5 AB and DC are two parallel and equal straight lines, named in the same direction. Find a centre which will map AB onto:
  a BA   *b* CD, under a half turn.
  Can you suggest a way of mapping AB onto DC?

6 O is the centre of a rectangle ABCD which is 8 cm by 5 cm. The diagonal BOD is drawn and POQ is a straight line meeting AB at P and CD at Q. Shade the triangles OPB and OQD.
  If the whole of the shaded area is one fifth of the area of the rectangle, calculate the area of APOD.

7 Name all the quadrilaterals which have the following properties. (Choose from rectangle, square, rhombus, kite and parallelogram.)
  a Two pairs of opposite sides parallel.
  b At least one pair of sides equal and parallel.
  c All four sides equal.
  d All four sides equal and one angle a right angle.
  e Half turn symmetry and one angle a right angle.
  f Half turn symmetry and two adjacent sides equal.
  g Only one diagonal an axis of symmetry.
  h Both diagonals axes of symmetry.

8 *a* Draw a parallelogram ABCD and mark a point E: (*1*) on a diagonal (*2*) inside ABCD, but not on a diagonal (*3*) outside ABCD.
  b Draw the image of ABCD under reflection in E in each case.
  c How many parallelograms do you see in each completed figure?

9 Explain, with the aid of a diagram, how you could use a tiling of parallelograms to identify points by 'coordinates' on a plane instead of using squared paper.

10 In each of the Figures 10(i) and 10(ii) the lines marked with arrows are parallel. Sketch these figures and write in the sizes of each of the other angles in the figures.

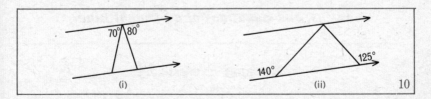
(i)   (ii)   10

Geometry

11 In Figure 11, PQRV is a kite. ∠VPQ = 120° and angles PVR and PQR are right angles. VQ is parallel to ST. Calculate the sizes of all the angles in the figure.

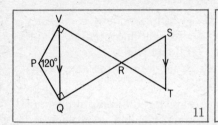

11

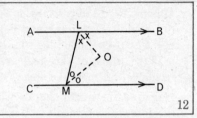

12

12 In Figure 12, AB and CD are parallel. LO bisects ∠BLM and MO bisects ∠DML.

a If ∠ALM = 68°, calculate the size of each of the other angles in the figure, including ∠LOM.

b Repeat *a* when ∠ALM = 42°.

c If ∠BLM = $2x°$ and ∠DML = $2y°$, write down a mathematical sentence about x and y.

What must be the size of ∠LOM in every case?

13 ABCD is a parallelogram and CDEF is a rectangle with E and F on the line of AB. Suggest a method of mapping △AED onto △BFC. Hence show that the area of ABCD is equal to the area of EFCD.
Deduce the formula for the area of a parallelogram.

14 The coordinates of the vertices of a parallelogram are $(p, 1)$, $(q, 1)$, $(p+2, r)$ and $(q+2, r)$. Calculate the area of the parallelogram in terms of p, q and r. (Assume that $q > p$ and $r > 1$.)

Revision Exercises on Chapter 3
Locus, and Equations of a Straight Line

Revision Exercise 3A

1 Show on separate diagrams the following sets of points:

a $\{(x, y): y = 2\}$ *b* $\{(x, y): y > 2\}$
c $\{(x, y): y = 5\}$ *d* $\{(x, y): 3 < y < 4\}$

Revision Exercises on Chapter 3

2 Sketch the following loci on the same diagram:
  a $\{(x, y): x = 2\}$ b $\{(x, y): x = 0\}$ c $\{(x, y): x = -3\}$
3 Show the following loci on separate diagrams:
  a $\{(x, y): 2 < x < 4\}$ b $\{(x, y): 2 \leqslant x \leqslant 6\}$
4 Use set notation to define the locus of points in the shaded regions of Figure 13.

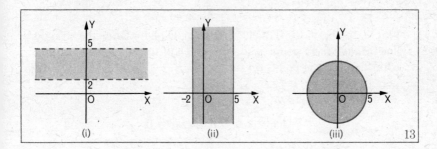

5 Using compasses or protractor or setsquare, draw an equilateral triangle ABC with sides 8 cm long. Show the sets of points inside the triangle which are:
  a 5 cm from A   b 5 cm from B   c 5 cm from C.
  Describe the intersections of pairs of these sets.
6 Draw an equilateral triangle ABC with sides 8 cm long. Show, by shading, the sets of points inside the triangle which are:
  a more than 5 cm from BC   b more than 5 cm from CA
  c more than 5 cm from AB.
  Describe the intersection of these three sets.
7 On the same diagram draw the sets of points, $A = \{(x, y): x = 5\}$, $B = \{(x, y): y = 3\}$, and describe the set $A \cap B$.
8 Describe in words the loci shown in colour in Figure 14.

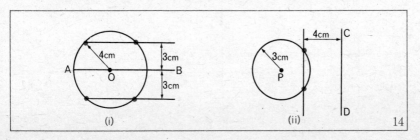

Geometry

9 Draw the path of the centre of a wheel of radius 2 cm which rolls round the inside of an equilateral triangle with sides of length 12 cm.

10 With axes OX and OY on squared paper, draw a diagram to illustrate the sets of points given by the following equations:

a $y = 3x$ *b* $y = 3x+3$ *c* $y = 3x-2$

11 Draw a diagram as in question *10* to illustrate the sets of points given by the following equations:

a $y = \frac{1}{2}x$ *b* $y = \frac{1}{2}x+3$ *c* $y = \frac{1}{2}x-2$.

12 Plot the points A (0, 4), B (1, 3), C (2, 2), D (4, 0). Draw the straight line through these points, and state which of the following equations correspond to the points on this line:

a $y = 4-x$ *b* $x+y = 4$ *c* $y = x+4$

Revision Exercise 3B

1 Show on separate diagrams the following sets of points:

a $\{(x, y): 0 < y < 2\}$ *b* $\{(x, y): y \geqslant -2\}$
c $\{(x, y): x = 2, y = 3\}$ *d* $\{(x, y): -1 < y < 0\}$

2 Sketch the following loci on the same diagram:

a $A = \{(x, y): 3 \leqslant x \leqslant 5\}$ *b* $B = \{(x, y): -1 \leqslant y \leqslant 2\}$

Describe in set notation and in words $A \cap B$.

3 Sketch the following loci on the same diagram:

$A = \{(x, y): y = 4\}$ $B = \{(x, y): 0 < x \leqslant 3\}$

Describe in set notation and in words $A \cap B$.

4 Use set notation to define the locus of points in the coloured regions of Figure 15.

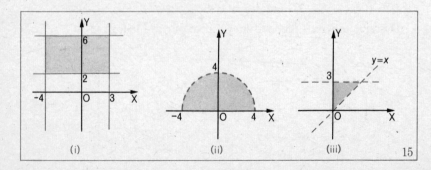

15

Revision Exercises on Chapter 3

5 Draw an equilateral triangle ABC with sides 8 cm long. Show, by shading, the sets of points inside the triangle which are:
 a less than 5 cm from A   *b* less than 5 cm from B
 c less than 5 cm from C.
 Describe in words the set of points which is the intersection of these three sets.

6 Draw a straight line AB of length 8 cm. Draw the locus of points 3 cm from A, and also the locus of points 5 cm from B. Hence find the set of points 3 cm from A and 5 cm from B.

7 Show on a diagram the set $A \cap B$ where
$A = \{(x, y): y > x+2\}$ and $B = \{(x, y): y < x+5\}$.

8 Describe in words the loci shown in colour in Figure 16.

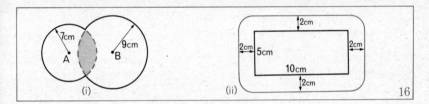

9 Draw the lines with equations:
 a $y = x+1$ *b* $y = x+3$ *c* $y = -x+1$ *d* $y = -x-1$

10 On the same Cartesian diagram draw lines which show the following sets of points:
 a $\{(x, y): y = x\}$ *b* $\{(x, y): y = 3x\}$ *c* $\{(x, y): 2y = x\}$

11 Sketch the lines with equations:
 a $2x+3y = 12$ *b* $3x-y = 6$ *c* $4x-5y+20 = 0$

12 Describe in words the locus in space of all points:
 a 5 cm from a fixed point O   *b* 5 cm from a fixed line
 c 5 cm from a fixed plane.

Arithmetic

Arithmetic

Note to the Teacher on Chapter 1

Two main themes are built into this chapter. One is the use of money in a variety of familiar household situations. Further aspects of 'social arithmetic' are studied later in the course. The other is the use and application of percentages. In this connection it is strongly recommended that a calculation such as $4\frac{1}{4}\%$ of £298 be set out as shown here.

This method is used throughout the course, and in this chapter in particular in the work on Discount (*Section* 1(iv)) and Interest (*Section* 2(iii)).

$$
\begin{array}{rl}
1\% \text{ of } £298 = & £2\cdot98 \\
& 4 \\ \hline
4\% \text{ ,, ,,} = & 11\cdot92 \\
\frac{1}{4}\% \text{ ,, ,,} = & 0\cdot745 \\ \hline
4\frac{1}{4}\% \text{ ,, ,,} = & £12\cdot66(5)
\end{array}
$$

In general, neat and orderly setting down of calculations should be stressed.

Section 1: It is not necessary to do both (i) and (ii), but the calculations in each are relatively straightforward, and the content is relevant. In part (iii), a few multiplications by 10 and by 100 have been included.

Section 2: No pupil in modern times should leave school without knowing something about the services offered by banks. Banks are only too pleased to supply brochures and, in some cases, wall charts which explain how banks can help their customers, and every classroom should have some of this publicity material available. Pupils should explore what services are offered, how to write cheques and how money can be saved—or borrowed!

Savings accounts lead naturally to the topic of Simple Interest. Note that this topic is introduced in stages, first calculating the interest on a whole number of pounds for a year, then proceeding to find interest for periods of less than one year. Calculation of simple interest for periods of more than a year is of course completely unrealistic.

It might be pointed out that bankers use ready reckoners to calculate interest (working to considerable numbers of significant figures), and also that the book idea of calculating in months is not in fact common practice but is done here only to avoid the tedious calculations involved in calculating by days (see Exercise 7, question

12). Pupils will appreciate why all banks now use computers to store and process data.

Unreal reverse questions involving finding the principal from a given Interest, Rate and Time, have no place in this course. But calculation of a rate per cent per annum is important, as it provides a means of comparison in the investment of money, and it gives the opportunity for practice in finding the percentage that one quantity is of another. Pupils should be encouraged to decide for themselves a suitable degree of approximation in answers; in particular they should realize that the only common fractions which should appear in answers are halves, thirds and quarters, and that all other fractions are best expressed as decimals rounded off to a suitable degree of approximation.

Section 3: Here again it is seen how the two fundamental calculations are applied to business. The artificial problem of finding a cost price from a given selling price and percentage profit has been avoided.

Section 4: It is necessary to show that applications of percentages are not restricted to money. The equivalence of common fractions, decimal fractions and percentages must be emphasized, and the chance is taken here to stress the meaning of 100%.

Social Arithmetic

1 Money in the home

(i) *The electricity account*

NATIONAL ELECTRICITY BOARD			
NAME ADDRESS	Period of account from 14 DEC 1971 to 15 MAR 1972		20 MAR 1972
Meter reading		Details of charges	£
Present	Previous		
15961	14295	DOMESTIC 93 @ 2·9p 1573 @ 0·8p	15·27

Figure 1 shows a typical electricity account. The meter readings are taken from the meter in the house. The unit of electricity is the kilowatt-hour. Details of charges (the *tariff*) vary from area to area, and may depend on the 'number of chargeable rooms' in the house. In the above account, the first 93 units have been charged at 2·9p each, and the rest at 0·8p each.

Example. Check the account in Figure 1.

Units used

```
   15 961          93 × 2·9p  = £  2·69          93        1573
 − 14 295        1573 × 0·8p  =   12·58         2·9         0·8
   ──────                                       ───       ─────
    1 666                     Total = £15·27    83 7       1258·4
                                                186 0
                                                ─────
                                                269·7
```

Arithmetic

Exercise 1

Calculate the total charge for electricity in each of the following:

	Present reading	Previous reading	Initial charge (*per unit*)	Remaining charge (*per unit*)
1	16000	15900	3p for first 60 units	1p
2	5456	3236	2p for first 100 units	0·5p
3	9471	8029	3·2p for first 80 units	1p
4	17536	14246	2·9p for first 90 units	0·8p
5	18514	15201	3·1p for first 120 units	1·1p
6	11111	9226	2·75p for first 80 units	1·1p

(ii) *The gas account*

Meter number	Tariff	Present reading	Previous reading	Consumption 100's cubic ft.	Calorific value	Therms	Amount
ZYX01	2	7154	6429	725	475	344·37 @ 9·5p	32·71
						Standing charge	3·75
						Total	36·46

National Gas Board   NAME ADDRESS   Date of issue 22·03·72   Period of account from 18·12·71 to 17·03·72

Note. The number of therms = consumption of gas in 100s of cubic feet × calorific value ÷ 1000

$$= \frac{725 \times 475}{1000} \text{ in the above account}$$

$$= 344\cdot 37$$

Example. Check the cash calculations in the account in Figure 2.

Cost of gas = 344·37 × 9·5p
 = £32·71
Standing charge = 3·75
Total = £36·46

```
     344·37
        9·5
     ------
     172 185
    3099 330
     ------
    3271·515
```

Money in the home

Exercise 2

Calculate the amount due for the following gas accounts.

	Number of therms	Cost per therm (p)	Standing charge (£)
1	100	9·5	3·75
2	250	8	3·75
3	24	9·5	4·25
4	440	11	2·50

5 Calculate the cost of gas when the calorific value is 480, 1 therm costs 12p, and the meter readings (in 100s of cubic feet) are as follows:

Present reading	5750	1192	23401
Previous reading	5450	642	18176

Project

Note the reading on the gas or electricity meter in your house once a week for several weeks. Show the readings on a suitable graph. Calculate the amount of gas or electricity used, and its cost.

(iii) *Ready reckoners*

If we buy 19 litres of petrol costing 7·5p per litre at a filling station the pump attendant does not do a multiplication sum to find the total cost. If the answer is not automatically registered on the pump he will use a *ready reckoner* which may look like this:

Litres	Two-star	Three-star	Four-star	Five-star
1	7p	7p	8p	8p
2	14	14	15	16
3	21	22	23	24
4	28	29	30	32
5	35	36	38	40
6	42	43	45	48
7	49	50	53	56
8	56	58	60	64
9	63	65	68	72
10	70	72	75	80
20	£1·40	£1·44	£1·50	£1·60
30	2·10	2·16	2·25	2·40
40	2·80	2·88	3·00	3·20
50	3·50	3·60	3·75	4·00

Arithmetic

Exercise 3

1. *Use the ready reckoner* on page 149 to find the cost of the following petrol:

Number of litres	8	9	30	50
Grade of petrol	Four-star	Three-star	Five-star	Two-star

2. *Use the ready reckoner* to find the cost of the following petrol:
   - a 11 litres of Two-star
   - b 15 litres of Three-star
   - c 60 litres of Four-star
   - d 38 litres of Five-star

3. *Use the ready reckoner* to help you to find the cost of:
   - a 7 litres, 70 litres and 700 litres of Two-star petrol
   - b 9 litres, 90 litres and 900 litres of Five-star petrol.

4. If a motorist has only 50p in his pocket, how many litres of each grade of petrol can he buy, and what change would he receive in each case?

5. Repeat question *4* for a motorist with only £1 to spend on petrol.

6. Make a ready reckoner for the cost of 5, 10, 15, 20, 25, 50, 100, 200, 300, 400 school lunches costing 15p each.

   Use the ready reckoner to write down the cost of:
   - a 25 lunches
   - b 75 lunches
   - c 150 lunches
   - d 250 lunches
   - e 880 lunches
   - f 615 lunches

7. The bus fare from Oldtown to Newtown is $9\frac{1}{2}$p. Make a ready reckoner to show the money collected for 1, 2, 3, ..., 10, 20, 30, 40, 50 fares.

   Use the ready reckoner to write down the sum collected for:
   - a 30 fares
   - b 36 fares
   - c 45 fares
   - d 90 fares
   - e 100 fares
   - f 500 fares
   - g 5000 fares
   - h 129 fares

(iv) Discount

A *discount* is a reduction in price, often given at a sale, or a special purchase of goods. It is usually given in the form 'a 5% discount', or '5p in the £ discount'.

Example. Find the total for the following bill, allowing a discount of $7\frac{1}{2}$%. 15 rose bushes at 27p each, 12 shrubs at $52\frac{1}{2}$p each, 7 trees at £1·25 each.

Money in the home

```
Bushes: 15 × 27p   =   £4·05           1% of £19·10 = £0·191
Shrubs: 12 × 52½p  =   £6·30                              7
Trees:  7 × £1·25  =    8·75
                       —————           7%  „    „    = 1·337
                       19·10           ½%  „    „    = 0·095
           Discount =   1·43                          ———————
                       —————                          £1·432
         Total cost = £17·67
```

Exercise 4

1 At their annual sale a firm allows 20% off all prices. What would you pay for:

   a a suit priced at £10   *b* a skirt costing £2·50
   c trousers at £3·65   *d* stockings at 27½p?

2 Calculate the actual cost of each of the following, after allowing the discount shown:

Marked price	£100	£15	£80	£36	60p	95p
Discount	25%	10%	5%	33⅓%	15%	20%

3 Find the total for the following bill, allowing 5% discount for cash payment: 10 rose bushes at 23p each, 2 standard roses at 39p each, 4 dozen raspberry canes at 24p per dozen, 6 apple trees at £1·10 each.

4 If 2½% discount is given on all goods sold in a sale, what would each of the following be reduced to?

   a a chair at £50   *b* curtains at £20   *c* a carpet at £31·60

5 A Cooperative store gives a 'dividend' of 7½p in the £ on all purchases. What dividend would customers receive if they spent:

   a £22   *b* £50   *c* £145   *d* £1000?

6 A firm uses 2500 litres of petrol per month, at a cost of 7·7p per litre less a discount of 3% for the large quantity. How much does the firm pay for the petrol?

7 A manufacturer offers cash registers at a discount of 33⅓%. How much has to be paid for machines originally priced at:

   a £48   *b* £84   *c* £200   *d* £145·50?

8 Find the cost of the following, allowing a discount of 2% on the total: 240 textbooks at 63p each, 100 slide rules at 87½p each, 30 dozen jotters at 5p per jotter.

| Arithmetic | 152 |

9 A fur coat is marked £199. What is the price at a sale when a discount of 35% is allowed?

10 A farm supplies milk to a factory at 11p per litre, and allows $2\frac{1}{2}\%$ discount for payment within one month. How much does the firm pay for 3750 litres, assuming prompt payment?

2 Money in the bank

We have all seen headlines like these in the newspapers. How can such losses be avoided? It is foolish to risk losing money by theft, fire or accident; the safe place for money is the Bank.

We are going to study two services that banks offer—current accounts and deposit accounts.

(i) *Current accounts (cheque accounts)*

To open a *current account*, you require to:
  deposit a sum of money in the bank. (The amount of money in your account at any time is called the *balance*);
  provide a reference (often your employer);
  provide a specimen signature.
Now instead of carrying cash you can pay by *cheque*.

Money in the bank

```
                                    23 May 1972    11-91-82
DB  Decimal Bank Limited
    1 HIGH STREET, ANYTOWN
Pay  Anytown Garage Ltd              or Order
Ten Pounds 08                       £10-08
                                    A.N.OTHER
                                    A.N.Other
"661316  11-91-82  C21421"
```

4

When you write a cheque you must:

  write the date;
  write the name of the person to whom the money is to be paid;
  write the sum of money in words;
  write the sum of money in figures in the box;
  sign the cheque.

Notes. *1* To avoid mistakes or fraud the decimal point is not used in writing £6·74; we write £6–74.

2 Banks do not deal in halfpence.

3 You must not sign a cheque for more than the balance in your account, otherwise the bank may not pay and you may be in serious trouble.

4 If you 'cross a cheque', by drawing two parallel lines on it, the cheque can only be paid through a bank to the correct person.

5 The bank will usually make a small charge for each transaction to cover costs.

Exercise 5

1 Sketch a cheque similar to the one in Figure 4, and complete the cheque for £12·25, payable to John Smith; include all the necessary details.

2 Repeat question *1* for another cheque, filling in all the information yourself.

3 There are several ways of crossing a cheque; find out about these.

Arithmetic

(ii) *Deposit accounts (savings accounts); interest*

You cannot use cheques with a *deposit account*, which is really an account in which your savings gain *interest*. Interest is expressed as a rate per cent per annum (p.a.).

Example. I deposit £450 in a Deposit Account which pays interest at the rate of $3\frac{1}{2}\%$ per annum. What interest do I get in one year?

$$\begin{array}{ll}
\text{Interest for 1 year} = 3\frac{1}{2}\% \text{ of £450} & \qquad 1\% \text{ of £450} = \text{£4·50} \\
\qquad \qquad \qquad \quad = \text{£15·75} & \qquad \qquad \qquad \qquad \quad \times 3 \\
& \qquad 3\% \quad ,, \quad ,, \quad = \text{13·50} \\
& \qquad \tfrac{1}{2}\% \quad ,, \quad ,, \quad = \text{2·25} \\
& \qquad \qquad \qquad \qquad \qquad \text{£15·75}
\end{array}$$

Exercise 6

Find the interest for 1 year on the money at the rates shown in questions *1–9*:

1   £500 at 3% p.a.   2   £160 at 5% p.a.   3   £33 at 7% p.a.
4   £200 at $3\frac{1}{2}$% p.a.   5   £12 at $3\frac{1}{2}$% p.a.   6   £1000 at $2\frac{1}{2}$% p.a.
7   £600 at $2\frac{1}{2}$% p.a.   8   £125 at 4% p.a.   9   £75 at 6% p.a.

10   A firm borrows £100000 from a bank. How much interest at $7\frac{1}{2}$% per annum must the firm give the bank at the end of one year?

11   The Government pays interest at the rate of $6\frac{1}{4}$% on Savings Certificates. How much interest must be paid each year for 21 million certificates costing £1 each?

5

Advertisements like the one in Figure 5 are meant to encourage people to save their money and to invest it in a bank, or a local authority, or with an insurance company or business firm.

The sum invested is called the *principal*, and interest is normally calculated each year (per annum, or p.a.) on complete pounds of the principal. The *principal + interest* is called the *amount*.

Money in the bank

Example 1. Find the amount of £125·83 after 1 year at 5% p.a.

Interest for 1 year = 5% of £125
                   = £ 6·25
Principal = £125·83
Amount = £132·08

1% of £125 = £1·25
                      5
5%   ,, = £6·25

Example 2. Calculate the interest on £23·50 at 6% p.a. for 8 months.

Interest for 1 year = 6% of £23
                   = £1·38
Interest for 8 months = $\frac{8}{12}$ × £1·38
                       = $\frac{2}{3}$ × £1·38
                       = £0·92

1% of £23 = £0·23
                     6
6% ,, = £1·38
                     2
                3)2·76
                 £0·92

Exercise 7

In questions *1–6* calculate the interest for 1 year on complete pounds of the principal. Hence give the amount of money after 1 year in each case.

1 £150·20 at 4%   2 £220·75 at 5%   3 £660 at 4%
4 £12·50 at 6%   5 £120 at 5½%   6 £55·38 at 2½%

In questions *7–12* calculate the interest, and then the amount.

7 £150 at 8% for 3 months   8 £1250 at 6% for 4 months
9 £27 at 8% for 9 months   10 £2350 at 2% for 6 months
11 £357·60 at 4¾% for 6 months   12 £75 at 8% for 100 days.

(iii) *Finding the rate of interest*

Example. A firm borrows £700, and clears the loan at the end of one year by paying £756. What rate per cent per annum had the firm to pay?

The principal is £700.

The amount is £756.

So the interest paid is £56. We have to find what percentage this is of the principal.

Rate % p.a. = $\frac{56}{700}$ × 100%

             = 8%

| Arithmetic |

Note. For interest on loans made to customers, banks usually charge about 8–14%, building societies 8–9%, hire purchase firms 15–30%.

Exercise 8

Calculate the rate per cent per annum interest in questions *1–8*.

1. Principal = £300; interest in 1 year = £24.
2. Principal = £64; interest in 1 year = £4.
3. Principal = £64; interest in 1 year = £4·80.
4. Principal = £256; interest in 1 year = £32.
5. Principal = £1250; *amount* at end of 1 year = £1300.
6. Principal = £56; *amount* at end of 1 year = £60·48.
7. Principal = £2; interest is 1p per week.
8. Principal = £36; interest is £1·62 every six months.
9. A Savings Certificate costs £1, and is worth £1·25 after 4 years. What rate per cent per annum simple interest does this represent?
10. A town council borrows £25000 at 7% p.a., and £15000 at 9% p.a. Express the total annual interest payable as a percentage of the total sum borrowed.

Topics to explore

1. Find out how many different kinds of bank there are in your town (Post Office Savings Bank, Joint Stock Banks, etc.).
2. Find out how Building Societies can look after your money.
3. Banks offer many services as well as keeping your money safe, and providing interest on it. Find out more about:
    Current accounts and deposit (savings) accounts.
    Special investment accounts.
    Giro transfers, and standing orders.
    Bank cards, and travel services.
    Loans, and overdrafts.

3 Money in business—profit and loss

(i) Actual profit and loss

A *shopkeeper* buys goods either direct from the *manufacturer* or through a *wholesaler*, and sells (*retails*) them to the *customer*.

If he sells his goods at a higher price than he paid for them then he makes a *profit* or *gain*.

If, for some reason, his selling price is less than his cost price then he sustains a *loss* on the transaction.

Thus   profit = selling price − cost price
and   loss = cost price − selling price

Example. A confectioner bought 72 Easter Eggs at $6\frac{1}{2}$ pence each. He sold half of them at $7\frac{1}{2}$ pence each and, after Easter, disposed of the rest at $4\frac{1}{2}$ pence each. Find his gain or loss on the deal.

$$\text{Cost price} = 72 \times 6\tfrac{1}{2}\text{p} = £4\cdot68$$
$$\text{Total selling price} = (36 \times 7\tfrac{1}{2})\text{p} + (36 \times 4\tfrac{1}{2})\text{p}$$
$$= 36 \times 12\text{p}$$
$$= £4\cdot32$$
$$\text{Loss} = £4\cdot68 - £4\cdot32 = 36\text{p}$$

Exercise 9

1   Calculate the profit or loss in each of the following:

Cost price	£12·50	£105	68p	£1·19	£2453
Selling price	£19·50	£100	86p	£1·19	£2616

2   A dealer bought 50 sheep at an auction sale for £8 each, and sold them privately at £12 each. Find his total profit.

3   A school shop bought 6 cartons, each containing 50 packets of crisps, for £13·50 altogether. If the packets were sold at 5p each, what was the total profit?

4   A stationer bought 150 Christmas decorations for £3·50. He sold as many as he could before Christmas at 4p each but was left with 20 which were then useless. What was his gain or loss on the deal?

5   A bookseller bought 4 dozen copies of a book at 42p each. He sold

Arithmetic

3 dozen copies at 60p per copy, and the rest at 30p each. Find his profit or loss.

6 A greengrocer bought a case containing 720 oranges for £10·80. He sold 300 of them at 3p each and 250 at 1½p each. The rest went bad and were thrown out. How much profit or loss did he make?

7 A florist bought 200 bulbs at 4½p each. 140 of them were classed as top size and sold at 6½p each. The remainder were put in bowls, three to a bowl, and sold at 52½p per bowl. If the bowls cost him 27½p each, and if the total cost of the bulb fibre was 65p, find his total profit.

(ii) *Percentage profit and loss*

For comparison purposes it is customary to express the actual profit or loss in the form of a percentage. This is sometimes taken as a percentage of the *cost price* although it is common trade practice for a shopkeeper to reckon his profit as a percentage of the *selling price*.

In all examples which follow you should assume that the gain or loss is to be expressed as a *percentage of the cost price unless it is otherwise stated*.

Example 1. An article is bought for 20 pence and sold for 25 pence. Express the profit as a percentage of:

a the cost price   *b* the selling price.

Actual profit = 25p − 20p = 5p

a Profit as a percentage of the cost price = $\frac{5}{20} \times 100\% = 25\%$

b Profit as a percentage of the selling price = $\frac{5}{25} \times 100\% = 20\%$

Example 2. Sugar bought at £2·80 per 50-kg bag is sold at 3½p per half kilogramme. Find the percentage profit.

$$\text{Cost price per 50 kg} = £2·80 = 280\text{p}$$
$$\text{Selling price per 50 kg} = 100 \times 3·5\text{p} = 350\text{p}$$
$$\text{Actual profit} = 70\text{p}$$
$$\text{Percentage profit} = \frac{\text{actual profit}}{\text{cost price}} \times 100\%$$
$$= \frac{70}{280} \times 100\%$$
$$= 25\% \text{ of the cost price}$$

Money in business—profit and loss

Exercise 10

Calculate the profit or loss as a percentage of the *cost price* in questions *1* to *4*.

1. Cost price = 30p; selling price = 36p.
2. Cost price = £15; selling price = £13·50.
3. Cost price = £8 per dozen; selling price = 75p each.
4. Cost price = 70p per kg; selling price = 14p per 100 g.

Calculate the profit or loss as a percentage of the *selling price* in questions *5* to *8*.

5. Cost price = £280; selling price = £350.
6. Cost price = 4½p; selling price = 4p.
7. Cost price = £2·16 per dozen; selling price = 16p each.
8. Cost price = £11 per 1000; selling price = 82½p per 50.
9. Articles cost £6 per dozen to make, and are sold for 60p each. Find the profit or loss as a percentage of: *a* the cost price *b* the selling price.
10. A dealer buys two motor bicycles for £200. He sells one for £112·50 and the other for £37·50. Calculate his loss as a percentage of: *a* the cost price *b* the selling price.
11. A dealer buys four cars for £250, £320, £375 and £455. He sells them for £300, £280, £460 and £640. Calculate his profit or loss on the whole deal as a percentage of the selling price.
12. A roll of floor covering 4 m broad and 40 m long was bought for £67·20 and sold at 49p per m². Find the gain or loss as a percentage of the cost price.

(iii) *To calculate the selling price*

Example. A bookseller bought 20 copies of a novel for £10·50 and priced them so as to gain $33\frac{1}{3}\%$ on his outlay. Find the marked selling price of each copy.

$$\begin{aligned}
\text{Profit} &= 33\tfrac{1}{3}\% \text{ of } £10·50 \\
&= \tfrac{1}{3} \text{ of } £10·50 \\
&= £3·50 \\
\text{Selling price} &= £10·50 + £3·50 = £14 \\
\text{Selling price per copy} &= £14 \div 20 = 70 \text{ pence}
\end{aligned}$$

Arithmetic

Exercise 11

Find the selling price in questions *1–5*, given that the profit or loss is reckoned as a percentage of the cost price.

1. Cost price £15; profit 10%
2. Cost price 65p; profit 20%
3. Cost price £1·20; loss 25%
4. Cost price 72p; profit $33\frac{1}{3}$%
5. Cost price £1·52; loss 75%
6. Articles bought for £6·48 per gross (144) are sold at a profit of $33\frac{1}{3}$%. Find the selling price of each.
7. A man bought a car for £840 and spent another £40 on extra fittings. Later he sold it at a loss of 35% on his total outlay. Find the selling price of the car.
8. An education authority requires 25 gross of a certain type of exercise book which costs the contractors 3p each to manufacture. If the contractor makes a profit of 10% find the cost of the exercise books to the authority.
9. Two different qualities of tea costing 28 pence and $32\frac{1}{2}$ pence per $\frac{1}{4}$ kg are blended in the ratio of 5 : 4. The mixture is sold at a profit of 25%. Find the selling price per $\frac{1}{4}$ kg.
10. Television sets are bought for £48 each and marked for sale to gain 20% on the cost price. Find the marked price. Most of the sets are sold at this price but when a new model is introduced the dealer is forced to clear the remainder at £40 each. Express the loss sustained by selling at this price as a percentage of
    - a his cost price
    - b his marked price (to nearest 1%)
11. A manufacturer makes 10000 toys of a certain type at a total cost of £450. Find the average cost of manufacturing one such toy.

    The manufacturer sells them to a wholesaler at a profit of $16\frac{2}{3}$% of his cost price. Find how much the wholesaler pays for 50.

    The wholesaler in turn sells them to a shopkeeper at 6p each. What is the wholesaler's profit reckoned as a percentage of what they cost him?

4 More percentages

(i) Common fractions, decimal fractions and percentages

A fraction can be expressed as a common fraction, a decimal fraction or a percentage.

For example, $\frac{3}{4} = 0.75 = 75\%$

The following results are worth remembering:

$10\% = \frac{1}{10} = 0.1$ $\qquad$ $33\frac{1}{3}\% = \frac{1}{3} = 0.33$ (rounded off to 2
$20\% = \frac{1}{5} = 0.2$ $\qquad\qquad\qquad\qquad\quad$ significant figures)
$25\% = \frac{1}{4} = 0.25$ $\qquad$ $66\frac{2}{3}\% = \frac{2}{3} = 0.67$ (rounded off to 2
$50\% = \frac{1}{2} = 0.5$ $\qquad\qquad\qquad\qquad\quad$ significant figures)
$75\% = \frac{3}{4} = 0.75$ $\qquad$ $100\% = 1$

Remember also that a percentage greater than 100 represents a number greater than 1.

For example, $125\% = 1.25$.

Exercise 12

Copy and complete this table:

	Common fraction	Decimal fraction	Percentage
1	$\frac{1}{4}$	...	...
2	...	0.5	...
3	...	...	20%
4	$\frac{1}{8}$	...	...
5	...	0.8	...
6	...	...	35%
7	$\frac{1}{10}$	...	...
8	...	0.7	...
9	...	...	150%
10	...	0.008	...

(ii) Using percentages

Example 1. Express $\frac{5}{6}$ as a percentage, rounded off to 1 decimal place.

$\frac{5}{6} = \frac{5}{6} \times 100\%$ $\qquad\qquad\qquad\qquad$ 6)500.00
$\phantom{\frac{5}{6}} = 83.3\%$ $\qquad\qquad\qquad\qquad\qquad\quad$ 83.33

Arithmetic

Example 2. In an election, 180 votes were cast for Mr A, 160 for Mrs B, and 80 for Miss C. What percentage of the total votes did Mr A receive?

$$\text{Total votes} = \begin{array}{r} 180 \\ 160 \\ 80 \\ \hline 420 \end{array}$$

$$\text{Mr A's percentage} = \frac{180}{420} \times 100\%$$
$$= \frac{1800}{42}\%$$
$$= 42 \cdot 9\%$$

Example 3. 24% of a certain number is 102. What is the number?

$$24\% \text{ of the number} = 102$$
so $$1\% \text{ of the number} = \frac{102}{24}$$
so $$100\% \text{ of the number} = \frac{102}{24} \times 100$$
$$= 425$$

Exercise 13

1. Express these fractions as percentages, to 1 decimal place if necessary:

   a $\frac{1}{3}$ b $\frac{3}{25}$ c $\frac{1}{6}$ d 0·6   e 0·085

2. A girl scored 114 out of 150 in English, 120 out of 160 in mathematics and 70 out of 80 in art. Find her percentage mark in each subject.

3. The rectangle shown in Figure 6 measures 10 cm by 2 cm. Write down its area. Calculate the percentage that each of the areas A, B, C, D is of the whole area.

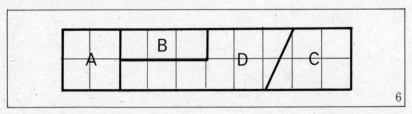

6

4. A school team plays 15 matches, winning 10 and drawing 4 of them. Allowing 2 points for a win and 1 point for a draw, express the points gained as a percentage of the total possible points.

More percentages

5. A man's hotel bill amounts to £12·70. What must he pay after adding a service charge of 10%?

6. Express the following as percentages:
   a. an increase of 12½p in the £
   b. a decrease of 9p in 45p
   c. a loss of 2 minutes in 1 hour
   d. a gain of 3 kg in 60 kg

7. Calculate: a. 15% of £1   b. 2½% of £1   c. 12½% of £5

8. 60 000 votes were cast in an election. Labour got 45% of the votes, Conservative got 32% and Liberal got the remainder. How many votes did each party get, and what was Labour's majority over Conservative?

9. A garden fertilizer consists of 50% sulphate of ammonia, 35% superphosphate and 15% sulphate of iron, by weight. What weight of each is in a 3-kg bag of fertilizer?

10. 80% of a number is 56. What is the number?

11. 110% of a number is 198. What is the number?

12. A boy scored 104 in an examination, and this was 65% of the possible score. What was the possible score?

13. 30% of a man's pay is deducted for income tax. If £7·35 is deducted for tax, what is his total pay?

14. A Building Society makes a loan of £2941 towards the cost of a house. If this is 85% of the value of the house, what is the house worth?

15. At one period the cost of living was increasing by 3% each year. If a household's weekly expenses were £9·40 in January one year, what would they be in January the following year? And in January the year after that?

Arithmetic

Summary

1. *Electricity and gas accounts* can be checked by calculation.

2. *Ready reckoners* are used to save time when certain calculations are repeated.

3. A *discount* of 20% on £50 = $\frac{1}{5} \times £50$ = £10;
   of $3\frac{1}{2}$% on £50 = £1·75.

1% of £50	=	£0·50
3% ,, ,,	=	1·50
$\frac{1}{2}$% ,, ,,	=	0·25
$3\frac{1}{2}$% ,, ,,	=	£1·75

4. *Banking.* Cheques can be used with current accounts. *Interest* is paid on deposit (savings) accounts. A *rate of interest* of 5% per annum gives interest of £5 in a year on a principal of £100.

5. *Profit and loss.* Percentage profit or loss may be calculated on the cost price or on the selling price.

6. *Percentages.* Some useful common fractions, decimal fractions and percentages:

   $10\% = \frac{1}{10} = 0·1$
   $20\% = \frac{1}{5} = 0·2$
   $25\% = \frac{1}{4} = 0·25$
   $50\% = \frac{1}{2} = 0·5$
   $75\% = \frac{3}{4} = 0·75$

   $33\frac{1}{3}\% = \frac{1}{3} = 0·33$ (to 2 sig. figs.)
   $66\frac{2}{3}\% = \frac{2}{3} = 0·67$ (to 2 sig. figs.)
   $100\% = 1$

Note to the Teacher on Chapter 2

Ratio was first introduced in Book 1 Arithmetic. The idea is revised in *Section* 1, and pupils encounter the technique of increasing or decreasing a quantity in a given ratio (which is equivalent to using a multiplying factor) in preparation for the methods of *Sections* 2 and 4.

The ratio method of dealing with problems in proportion, both direct (in *Section* 2) and inverse (in *Section* 4) is introduced by using the idea of a mapping. Pupils have met this in algebra and it will be used again in arithmetic when they study the Slide Rule and Logarithms. In a direct proportion connecting two variables we may write

$$\begin{array}{c|cc} \text{First variable} & a & b \\ & \updownarrow & \updownarrow \\ \text{Second variable} & p & q \end{array}$$

It is important to realize that the relation between the two variables can be expressed either in the form $\frac{a}{p} = \frac{b}{q}$ (in which we are dealing with a *rate* and are using a *unitary* method) or in the form $\frac{a}{b} = \frac{p}{q}$ (in which we use equal *ratios* and arrive at a *ratio* method). In the above example, if the quantities are connected by inverse proportion, we can express the relation either in the form $ap = bq$, a *product* method which is often by far the easiest and most obvious way of solving a problem (as in Example 1 in *Section* 4) or as $\frac{a}{b} = \frac{q}{p}$. This gives rise to a *ratio* method which is preferable in Example 2 on page 176 and in questions where the 'peculiar units' of the product cause difficulty.

The correct form of the multiplying factor, $\frac{p}{q}$ or $\frac{q}{p}$, can always be found by considering whether we are increasing or decreasing the given value. It will be noticed that the corresponding numbers of books and numbers of pence in the cost at the beginning of *Section* 2, and the numbers of kilometres and hours at the beginning of *Section* 4, have been written in a 'vertical' form and not in the more usual 'horizontal' form as is done earlier in this paragraph. There are two reasons for this. First, the form chosen is exactly that used when writing a mapping in algebra. Second, the form fits neatly the traditional way of writing 'proportion sums'. Pupils may have learned such a method in earlier years and they will have nothing to unlearn.

It is important that pupils should be careful about whether quantities are in proportion or not, and Exercise 4 is included for this specific purpose.

Maps and plans are amongst the commonest examples of proportion and will be studied again later in the geometry course in connection with similar figures. It must be realized that the use of the Representative Fraction, which was rather an unreal exercise in the British system of measures, must now be regarded as the standard method of stating a scale when the metric system is used. All metric maps state the scale as an R.F., and only occasionally does the map give the secondary form '1 cm represents 5 km'. Pupils will be constantly faced with the conversion km ↔ cm. As shown in the examples of *Section* 3 they should not try to memorize the conversion factor 100000, but should convert via metres, km ↔ m ↔ cm; thus 'multiply by 1000 (to give metres), and by 100 (to give centimetres), and the reverse.

Teachers should try to show their classes examples of scale plans and maps to illustrate this topic and make it more meaningful.

Section 5 introduces the principle that the graph connecting two quantities which are in direct proportion is a straight line through the origin. In Figure 4 the graph is, of course, a set of points corresponding to 1, 2, 3, ... books, and intermediate points have no meaning. In contrast, in question *2* of Exercise 7 meaning can be attached to intermediate points so the graph can properly be drawn as a continuous line; this important point may well be discussed with able pupils. Again, in the case of quantities in inverse proportion it may be profitable to discuss with able pupils the typical hyperbola form, and to point out that in a scientific experiment in which results produce a graph of this form, the quantities under investigation could possibly be inversely proportional. Note that the straight-line graph as a graph of direct proportion will be studied again in Chapter 4 of this book, in the special case of constant average speed, and is approached in a more general way in Chapter 3 on 'Locus' in geometry.

2 Ratio and Proportion

1 Ratio

In a football match, Rovers scored 6 goals and United scored 3 goals. We can compare their scores in two ways:

a Rovers scored 3 goals more than United. Here we are comparing the scores by finding their *difference*.

b Rovers scored twice as many goals as United. Here we are comparing the scores by finding their *ratio*, $\frac{6}{3}$, or $\frac{2}{1}$, or 2 : 1.

The ratio is often the most useful measure of comparison. It can only be used with two quantities of the same kind, and is a *number* (being a number of times).

Example 1. If one village has a population of 12 000, and another has a population of 8000, the ratio of the populations is $\frac{12000}{8000} = \frac{3}{2}$.

Example 2. If costs are *increased in the ratio* 10 : 9, what would a cost of £180 become?

$$\text{New cost} = £180 \times \tfrac{10}{9} = £200.$$

We could also have said that the costs are *multiplied by the factor* $\frac{10}{9}$.

Exercise 1

1 Express each of the following ratios in its simplest form:
 a 25 : 40   b 32 : 24   c $2\frac{1}{4} : 1\frac{1}{4}$
 d 18p : 45p   e 600 m : 1 km   f 2 kg : 800 g

2 Express each of the following ratios in its simplest form:
 a 50p to £2   b 120p to 80p   c 75 cm to 1 m
 d 1·5 cm to 9 mm   e 1 hour to 55 min   f $\frac{1}{2}$ mm to $\frac{3}{4}$ mm

Arithmetic

3 A football pitch is 108 metres long and 72 metres wide. What is the ratio of:
  a its length to its breadth   *b* its breadth to its length?

4 What is the ratio of the lengths of two sides of a square?

5 On squared paper draw a rectangle whose length and breadth are in the ratio 2 : 1. What have you *not* been told?

6 *a* If in question *5* the breadth is 5 cm, what is the length?
  b If the length is 15 cm, what is the breadth?

7 Two squares have sides of lengths 15 mm and 12 mm. Find the ratio of:
  a the lengths of their sides   *b* their perimeters   *c* their areas.

8 A man earns £25 per week and his wife earns £20. What is the ratio of:
  a the man's wage to his wife's   *b* the wife's wage to her husband's
  c the man's wage to the total household income?

9 The population of a town is 11 700, of whom 5200 are under 21 years of age. Calculate the ratio of the number under 21 to the number who are 21 or over.

10*a* Calculate: (*1*) $162 \times \frac{5}{6}$ (*2*) $162 \times \frac{11}{9}$. In each case state whether the number 162 has been increased or decreased.
  b Suppose you have to multiply a number by one of the following ratios. State whether the effect will be increase, decrease or no change.
  (*1*) $\frac{2}{3}$ (*2*) $\frac{5}{4}$ (*3*) $\frac{7}{7}$ (*4*) $\frac{18}{29}$ (*5*) $\frac{37}{33}$ (*6*) $\frac{143}{97}$ (*7*) $\frac{201}{202}$

11*a* Multiply £24 by the ratios: (*1*) $\frac{5}{3}$ (*2*) $\frac{7}{6}$ (*3*) $\frac{11}{8}$
  b Multiply 75 cm by the ratios: (*1*) $\frac{3}{5}$ (*2*) $\frac{2}{3}$ (*3*) $\frac{7}{15}$
  c Multiply $4\frac{1}{2}$ m by the factors: (*1*) $\frac{4}{3}$ (*2*) $\frac{7}{6}$ (*3*) 1
  d Multiply £2·52 by the factors: (*1*) $\frac{2}{3}$ (*2*) $\frac{5}{7}$ (*3*) $\frac{11}{12}$
  e Multiply 100 by the factor $\frac{11}{10}$, then multiply your answer by $\frac{9}{10}$.

12*a* Increase 140 in the ratio: (*1*) 8 : 7 (*2*) 7 : 5
  b Decrease 153 in the ratio: (*1*) 4 : 9 (*2*) 15 : 17

Direct proportion

2 Direct proportion

The costs of different numbers of a certain book are shown below.

Number of books		Number of pence in the cost
1	←————————→	30
2	←————————→	60
3	←————————→	90
4	←————————→	120
10	←————————→	x
n	←————————→	y

One-to-one correspondence. There is one, and only one, cost corresponding to a given number of books; and one, and only one, number of books corresponding to a given cost. So we say that there is a *one-to-one correspondence* between the number of books and the cost, as indicated by the arrows.

Rate. The *rate* giving the number of pence per book is the same for any pair of numbers in a row of the table.

From the first row we see that the *rate* is 30 pence per book.

The rate is also given by $\frac{60}{2}$, or $\frac{90}{3}$, or $\frac{120}{4}$, or $\frac{x}{10}$, or $\frac{y}{n}$.

Ratio. The ratio $\frac{\text{number of books in second row}}{\text{number of books in fourth row}} = \frac{2}{4} = \frac{1}{2}$

The ratio $\frac{\text{cost of books in second row}}{\text{cost of books in fourth row}} = \frac{60}{120} = \frac{1}{2}$

We can see from the table that corresponding *ratios* of numbers of books and costs are equal.

Thus $\frac{1}{3} = \frac{30}{90}$; $\frac{2}{3} = \frac{60}{90}$; $\frac{1}{10} = \frac{30}{x}$; $\frac{10}{n} = \frac{x}{y}$.

When the number of books is doubled, the cost is doubled; when one is halved, the other is halved. The number and the cost increase or decrease in the same ratio, and we say that the cost is *directly proportional* to the number of books.

Arithmetic

(i) *Unitary method of calculation*

Here we find the *rate*—the cost of 1 book, the time to travel 1 km, etc.

Example 1. If 5 metres of certain material cost £4·25, find the cost of 12 metres.

5 metres cost £4·25

so 1 metre costs $\dfrac{£4·25}{5}$ = £0·85

and 12 metres cost £0·85 × 12 = £10·20

Example 2. If I can travel 200 km in $2\frac{1}{2}$ hours on a motorway, how long would I take for 140 km in the same conditions?

200 km take 150 min
so 1 km takes $\frac{150}{200}$ min = $\frac{3}{4}$ min
and 140 km take $\frac{3}{4}$ min × 140 = 105 min
= 1 hour 45 minutes

Exercise 2

1   Find the *rate* in each of the following, in the units stated:
  a   8 oranges cost 24p;   pence per orange
  b   20 bars of chocolate cost 90p;   pence per bar
  c   273 units of electricity per week;   units per day
  d   20 km in 4 hours;   km per hour
  e   20 apples weigh 2 kg;   apples per kg
  f   140 km take $2\frac{1}{2}$ hours;   km per hour

2   A typist typed 900 words in 1 hour. What is her rate of typing in words per minute?

3   What is the rate of another typist who typed 1080 words in 1 hour?

4   I paid £26 for a year's television rental. What rate per week was charged?

5   I paid £208 for a year's rent of a house. What rate per week was charged?

6   6 metres of cloth cost £1·74. What is the cost of 1 metre? of 11 metres?

Direct proportion

7 A ticket for a journey of 24 km cost 72p. What is the cost for 1 km? for 60 km?

8 A car travels 456 km on 48 litres of petrol. How far should it go on 30 litres?

9 A man earned £5·76 for an 8-hour day. How much would he earn at the same rate for a 42-hour week?

10 A factory made 37500 cars in a year. Find the rate of production per week and per day, if the factory is closed for two weeks in the year and the men work a 5-day week.

11 A car travels at 20 metres per second. How many metres does it travel per hour? How many kilometres per hour?

(ii) *Ratio method of calculation*

Part of the table on page 167 is shown below:

Number of books		Number of pence in the cost
3	←———————→	90
4	←———————→	(120)

The ratio of the numbers of books is $\frac{4}{3}$, and the ratio of the corresponding costs is $\frac{120}{90} = \frac{4}{3}$.

Suppose the number in brackets is missing. We see that 4 can be got from 3 by multiplying 3 by the factor $\frac{4}{3}$. So we can find the missing number by multiplying 90 by $\frac{4}{3}$, giving 120.

This suggests a method of solving problems where quantities are in direct proportion.

Example 1. 27 cm³ of gold weigh 522 g. Find the weight of 18 cm³.

Number of cm³		Number of g
27	←———————→	522
18	←———————→	( )

The ratio of the second volume to the first is $\frac{18}{27}$, so the weight must also be multiplied by the factor $\frac{18}{27}$. So

$$18 \longleftrightarrow 522 \times \frac{18}{27}$$
$$= \frac{522}{1} \times \frac{2}{3}$$
$$= 348$$

The weight of the gold is 348 g.

Arithmetic

Example 2. Find the number of French francs you will receive in exchange for £12 if £5 buys 66 francs.

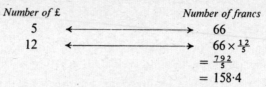

You will receive 158 francs.

Exercise 3A

1. Handkerchiefs are priced at 3 for 29p. Find the cost of 1 dozen.
2. Soap costs 66 pence for 8 cakes. What will you pay for 12 cakes?
3. Bulbs are sold at 75 for 60p. What should 100 cost?
4. A car travels 456 km on 40 litres of petrol. How far should it go on a full tank of 60 litres?
5. 72 books weigh 9 kg.
   - a. How many books weigh 6 kg?   b. What is the weight of 80 books?
6. 50 dollars are exchanged for £20·75. How much are 80 dollars worth in £s?
7. £1 buys 120 Belgian francs. What is the price in £s of a bus trip advertised at 105 francs?
8. A hotel charge for 5 days was £11·50. What would the charge be for 8 days?
9. A boy scored 56 out of 80 in a test. What percentage is this? (that is, what is his score out of 100?)
10. On a plan, 5 cm represents 8 m.
    - a. What actual length is represented by 9·5 cm?
    - b. What length on the plan represents 28 m?

Exercise 3B

1. Oranges cost 28 pence per dozen. Find the cost of 27 oranges.
2. 15 articles cost 52½p. Find the cost of 36 such articles.
3. 500 twopence coins placed in a row measure 13 metres. What would 75 coins measure?

Direct proportion

4 A firm's weekly wage bill for 350 employees is £6300. At the same average wage, what would the weekly wage bill be for 250 employees?

5 A stack of 350 sheets of paper is 2·1 cm high. How high would 500 sheets stand?

6 One week a licensed grocer bought 15 bottles of wine for £9·60. What would he pay the next week when he ordered two dozen similar bottles?

7 110 Indian rupees are worth £6. Find to the nearest rupee how many rupees you get for £5.

8 42 Swiss francs can be exchanged for £4. What are 104 Swiss francs worth in British money? (Answer to nearest p.)

9 A boy takes 150 steps in walking 120 metres. How far would he go in taking 250 steps? How many steps would he take in 100 metres? (to the nearest whole number).

10 When the rates were £0·95 in the £ a man paid £79·80 in rates on his house. What would he pay when the rate per £ increased by 15 pence?

* * *

Note. We must be sure that the quantities *are* in direct proportion before carrying out calculations like the ones above. Some of the questions in Exercise 4 contain quantities which are not in proportion.

Example 1. Henry 8th had 6 wives; how many wives had Henry 4th?

There is no connection at all between the number of the king and the number of his wives, so proportion does not apply.

Example 2. A satellite which travels in an orbit 1600 km above the earth takes 2 hours to circle the earth. At what height will a satellite be travelling if it takes 3 hours to circle the earth?

In this question you just don't know whether the height is directly proportional to the number of hours, so you cannot answer the question. (In fact the height is 4160 km, not 2400 km as it would be if the quantities were proportional.)

Arithmetic

Exercise 4

1 A seedsman's catalogue gives prices for daffodil bulbs as follows:

Number of bulbs	12	50	100
Cost in pence	20	75	140

   a Is the cost in direct proportion to the number of bulbs?
   b How much would you expect to pay for 6 bulbs? for 200 bulbs?

2 Television hire charges are as follows: 50-cm screen, 45p per week; 60-cm screen, 50p per week. By working out ratios, find if the hire charge is proportional to the dimension of the screen.

3 The ticket for a 24-km journey costs 54p. What should it cost for a 60-km journey?

4 To insure his house for £3000 a man has to pay £3·60 to an insurance company. What would he pay to insure it for £3500?

5 An orchestra takes 35 minutes to play Beethoven's Fourth Symphony. How long will it take to play his Sixth Symphony?

6 A building 20 m high casts a shadow 9 m long. What length of shadow would be cast at the same time by a tree 15 m high?

7 A shopkeeper can buy 48 toys for £36·40. What should he pay for 120 similar toys?

8 A man with 3 children earns £1800 per year. What does a man with 5 children earn?

9 At school parties 7 cakes are provided for every 5 children. How many cakes should be provided for 96 children?

10 A man owning 1600 shares in a company received an income of £60. Another man received £48; how many shares had he?

11 An agent received commission of £35 on sales totalling £1400. At the same rate
   a what commission would he receive on sales of £6200?
   b what sales would earn him a commission of £24?

12a In a book store 100 books occupy 1·80 m of shelf. What length of shelf is required for 540 similar books?
   b If shelves are 1·25 m long, how many shelves are needed for 540 books?

3 Maps and plans

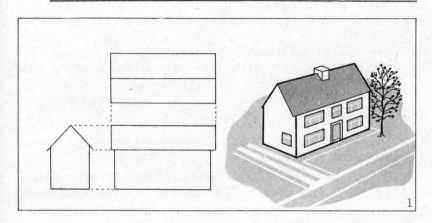

Maps of countries, and plans of buildings and cars and aircraft, are based on direct proportion. The ratios of distances on the map and corresponding distances on the ground are equal.

The *scale* of the map or plan is often shown by means of *a marked line* as in Figure 2.

Here 1 cm represents 1 km, and 0·1 cm represents 100 m.

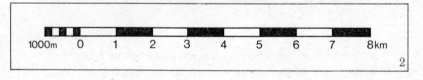

The scale of a map is sometimes given by the actual *ratio*:

$$\frac{\text{distance on the map}}{\text{corresponding distance on the ground}}$$

This ratio is called the Representative Fraction (R.F.) of the map.

For the scale shown in Figure 2, the R.F. is

$$\frac{1 \text{ cm}}{1 \text{ km}} = \frac{1 \text{ cm}}{1000 \text{ m}} = \frac{1 \text{ cm}}{1000 \times 100 \text{ cm}} = \frac{1}{100000}$$

This is usually written in the form 1/100000, or 1 : 100000.

Arithmetic

Example. On a map with R.F. 1 : 10000, calculate:
- *a* the distance on the map representing 1 km on the ground
- *b* the distance on the ground represented by 12·5 cm on the map.

a Distance on map = $\frac{1}{10000}$ of distance on ground
= $\frac{1}{10000}$ × 1 km = $\frac{1}{10000}$ × 1000 m = $\frac{1}{10000}$ × 1000 × 100 cm
= 10 cm

b Distance on ground = 10000 × distance on map
= 10000 × 12·5 cm = 125000 cm = 1250 m = 1·25 km
(The answer to *b* could have been found from *a*.)

Exercise 5

1 A plan has a scale of 1 : 100. What distance on the plan represents:
a 154 cm   *b* 6320 cm   *c* 26 cm?

2 On a map with a scale of 1 : 1000 what actual distance is represented on the map by: *a* 12 cm   *b* 1·34 cm   *c* 0·285 cm?

3 The scale of a plan is given in Figure 3.
a By measuring, find the distance which is represented by 1 cm on the plan.
b What is the R.F. of the plan?
c What distance is represented by a line of length 3·65 cm on the plan?
d What length on the plan would represent an actual length of 27·8 m?

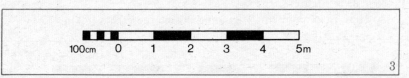

3

4 *a* What is the R.F. for a map with a scale of 1 cm to 2 km?
b If 2 towns are 94 km apart, what is the distance between them on the map?
c If the distance between another two towns on the map is 27·4 cm, how far apart are they?

5 The R.F. of a map is 1/50000. Calculate:
a what distance on the map represents 42 km
b how many km are represented by 11·8 cm.

6 *a* On a map of R.F. 1/200000 what is the actual distance between two towns whose distance apart on the map is 3·8 cm?
b If two hills are 14·8 km apart, what is the distance between them on the map?

Inverse proportion

7 The scale of a building site plan is 1/500.
  a If a rectangular plot on the plan measures 6·8 cm by 3·7 cm, what are the actual measurements? What is the real area of the plot?
  b What area in m² does 1 cm² represent? Hence find the real area of a building which occupies 32 cm² on the plan.

8 The distance from Rome to Milan is 612 km. How far apart would these cities appear on a map with R.F.:
  a 1/1 000 000   *b* 1/750 000?

4 Inverse proportion

The times taken by aircraft to cross the Atlantic at various speeds are shown below.

Number of km per hour		Number of hours taken
480	⟵——————⟶	10
600	⟵——————⟶	8
800	⟵——————⟶	6
960	⟵——————⟶	5
1200	⟵——————⟶	x
a	⟵——————⟶	y

One-to-one correspondence. The table shows a one-to-one correspondence between the number of km per hour and the number of hours taken.

Products. The *product* of the number of km and the number of hours is the same for any row in the table, and gives the distance of the Atlantic crossing, 4800 km.

Thus $480 \times 10 = 600 \times 8 = 800 \times 6 = 960 \times 5 = 1200x = ay = 4800$.

Ratio. The ratio $\dfrac{\text{speed in first row}}{\text{speed in third row}} = \dfrac{480}{800} = \dfrac{3}{5}$

The ratio $\dfrac{\text{number of hours in first row}}{\text{number of hours in third row}} = \dfrac{10}{6} = \dfrac{5}{3}$, the *reciprocal*, or *multiplicative inverse*, of $\tfrac{3}{5}$.

We can see from the table that corresponding ratios are reciprocals of each other

$\dfrac{480}{600} = \dfrac{8}{10}$; $\dfrac{800}{960} = \dfrac{5}{6}$; $\dfrac{960}{1200} = \dfrac{x}{5}$; $\dfrac{1200}{a} = \dfrac{y}{x}$, and so on.

Arithmetic

When the speed is doubled, the time is halved; when the speed is halved, the time is doubled. Since one ratio is the *inverse* of the corresponding ratio, we say that the speed is *inversely proportional* to the time.

(i) *Product method of calculation*

Example. If a box of sweets is divided among 15 children, they get 12 sweets each. How many sweets will each child get if the sweets are divided among 20 children?

The total number of sweets in the box $= 15 \times 12$ sweets $= 180$ sweets.

So if the box is divided among 20 children, each gets $\frac{180}{20}$ sweets, i.e. 9 sweets.

In this method we sometimes encounter peculiar units. For example, if it takes 10 hours for 3 men working together to re-spray a motor-car, we must talk about 30 man-hours. If you burn a 2-kilowatt electric fire for 8 hours, you use 16 kilowatt-hours of electricity (and a kilowatt-hour is usually called a *unit*). Another unit of this kind is passenger-kilometre.

(ii) *Ratio method of calculation*

Example 1. Find what average speed must be kept up to travel from Glasgow to Penzance by car in 12 hours, given that it takes 16 hours to make the journey at an average speed of 57 km per hour.

Number of hours		Number of km per hour
16	⟵————————⟶	57
12	⟵————————⟶	( )

Now the time has been changed by a factor of $\frac{12}{16}$ and we know that the speed required is inversely proportional to the number of hours taken, so the speed must be changed by a factor of $\frac{16}{12}$. The second pair of corresponding values will be written:

$$12 \longleftarrow\longrightarrow \quad 57 \times \tfrac{16}{12}$$
$$= 57 \times \tfrac{4}{3}$$
$$= 76$$

The average speed required is 76 km/h.

Example 2. BBC Radio 4 is broadcast on a wavelength of 371 metres and has a frequency of 809 kilocycles per second. What is

Inverse proportion

the frequency of Radio 2, which has a wavelength of 1500 metres, given that frequency is inversely proportional to wavelength?

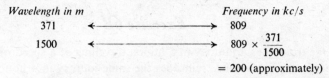

$$= 200 \text{ (approximately)}$$

The frequency is 200 kilocycles per second approximately.

Exercise 6A

1 Which of the following are in inverse proportion?
a The number of ice-creams you buy *and* their total cost.
b The number of men on a job *and* the time taken to finish the job.
c The speed of a moon probe *and* the time taken to reach the moon.
d A boy's age *and* his mark in mathematics.
e The number of cattle in a field *and* the time taken to eat the available grass.
f The number of ice lollies you can buy *and* the cost per ice lolly, given you have only a fixed sum available.

Work out questions *2–8* as examples of inverse proportion:

2 A boy knows that when he cycles a certain distance in 40 minutes his average speed is 15 km per hour. If he reduces his time to 30 minutes, what is his average speed?

3 In how many days could 20 men do a piece of work which 25 men can do in 32 days?

4 A car takes 12 hours for a journey at an average speed of 50 km/h. What average speed would be necessary to cover the journey in 10 hours?

5 A teacher has enough money to order 48 paperback books at 60p each. If he decides instead to order books with stiff covers at 90p each, how many can he get for the same total cost?

6 A contractor estimated that he could do a certain job in 9 months with 280 men. If he is asked instead to do the work in 7 months, how many *more* men would he need to employ?

7 A camp commandant estimates that he has enough food for 6 meals for 150 boys. If 30 more boys arrive than are expected, how many meals can he supply to each boy?

Arithmetic

8   A farmer has enough feeding stuff for his 50 cattle to last 10 weeks. If he sells 10 of his cattle, how much longer will the feeding stuff last?

Exercise 6B

1   One train does a journey in 5 hours, travelling at a speed of 56 km per hour. If a second train does the same journey in 4 hours, what is its speed?

2   A contractor estimated that he could do a job in 11 months with 96 men. If he is asked instead to do the job in 8 months, how many more men would he need to employ?

3   A space probe travelling at 9600 km per hour takes 40 hours to reach the moon. How long would it take if it travelled at 40 000 km per hour? (Assume that the earth and moon stay about the same distance apart; what distance?)

4   A class annually raises money to hire a bus for an outing. One year the cost of the bus per km was 3p, and the class had just enough money to go on a 120-km outing. The following year the cost per km was 4p. If the class raised the same amount of money, how far can they travel the second year?

5   A man walks for $6\frac{1}{2}$ hours at 6 km per hour. Had he walked at $6\frac{1}{2}$ km per hour, for how long would he have had to walk to cover the same distance?

6   A camp commandant estimates that he has enough food for 8 main meals for 140 boys. If 84 more boys arrive than are expected, how many main meals can he supply to each boy?

7   It is known that, for a given voltage, the current in a circuit is inversely proportional to the resistance in the circuit. When the current is 2 amperes the resistance is 3 ohms.
  a Find the resistances for currents of 5, $1\frac{1}{2}$ and 0·2 amperes.
  b Find the currents for resistances of 5, 2 and 12 ohms.

8   It is known that, for a given mass of gas, the volume is inversely proportional to the pressure. When the volume is 60 cm^3 the pressure is 1·5 atmospheres.
  a Find the volumes for pressures of 1, 2, 0·8 atmospheres.
  b Find the pressures for volumes of 90, 48 and 200 cm^3.

5 Graphs

Here is a table of costs of the books referred to on page 167.

Number of books	0	1	2	3	4	5	6
Number of pence in the cost	0	30	60	90	120	150	180

Figure 4 shows a graph relating the number of books to their cost.
You will find that when two quantities are related by direct proportion the graph connecting them consists of a set of points which may be joined by a straight line passing through the origin. (You may be able to visualize the idea of the straight line from the equal ratios—'double the distance along, double the distance up', etc.)

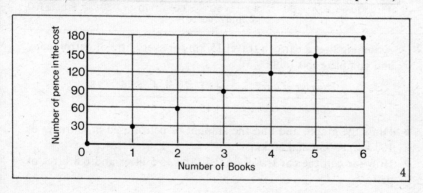

4

Exercise 7

1   Figure 5 shows a graph for converting Austrian schillings into British pence, and vice versa. Read off, as accurately as you can, the number of:

a British pence that can be obtained for 12, 30, 45 and 56 Austrian schillings

b Austrian schillings that can be obtained for 25p, 35p, 72p and 90p.

Arithmetic

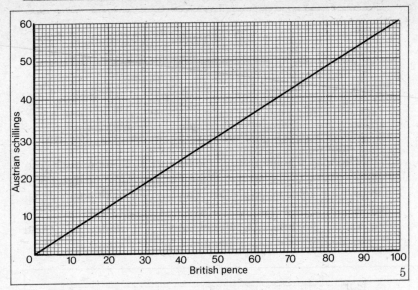

2 Assuming that a car can travel 15 km for every litre of petrol, copy and complete this table:

Number of litres	1	2	3	4	5	6	7	8
Number of kilometres	15							

a Draw the graph, and find the amount of petrol used in journeys of 100 km, 85 km and 50 km.

b How far can the car travel on 2·5 litres, 5·5 litres and 6·8 litres of petrol?

3 The premium charged by an insurance company is directly proportional to the amount insured. Copy and complete this table:

Amount insured (£)	400	800	1200	1600	2000	2400
Premium (£)	1					

Draw a graph, and hence find the approximate premium to insure amounts of £1500, £1800, £2100 and £2250.

4 A manufacturer of weedkiller recommends that 1 kg chemical should be used for every 50 m² of ground. Copy and complete this table:

Area of ground in m^2	50	100	150	200	250	300
Weight of chemical in kg	1					

Miscellaneous questions 181

   a Draw a graph and find the approximate weight of chemical to cover: 80 m², 120 m², 220 m² and 275 m² of ground.
   b How much ground can be covered by 1·5 kg, 2·5 kg and 4·5 kg chemical?

5 The distance from the earth to the moon is about 384 000 km. A rocket travelling at 9600 km/h will take $\dfrac{384\,000}{9600}$ hours, i.e. 40 hours, to travel this distance. Copy and complete this table by calculating each time of travel.

Speed of rocket in km/h	9600	12800	16000	19200	25600	32000
Time of travel in hours	40					

The graph connecting these quantities is not a straight line. Why not?

Draw a smooth graph, and from it estimate the time taken by rockets travelling at 10000 km/h and 30000 km/h.

When two quantities are related by *inverse proportion* the graph is *not* a straight line, but a curve of the shape you have found.

6 If the rate of exchange is 18 Danish kroner to £1, draw an appropriate graph. Hence find the number of:
   a pence that can be obtained for 10, 15 and 16 kroner
   b kroner that can be obtained for 20p, 55p and 72p.

---

6 Miscellaneous questions

Exercise 8A

1 A cook finds that in order to give 100 pupils lunch she must order 30 kg of potatoes. What quantity must she order for 350 pupils?

2 When going on holiday to France, Mary changed £7 to francs, and received 91 francs. How many francs would Alice be given for £5?

3 A boy can put 400 leaflets into envelopes in 30 minutes. How long should it take him, at the same rate, to put 500 leaflets into envelopes (to the next whole number of minutes)?

4 A boy's average step is 60 cm, and his father's is 72 cm. The boy

Arithmetic

takes 840 steps for a certain distance. How many steps would his father take?

5 Five persons won £154 each in a competition. Two others claimed that they should share the prize. How much would each person receive in this case?

6 A record played at 45 rpm takes 13 minutes. How long would it take if played at 78 rpm?

7 A fitted carpet for a room 4 m by 3 m costs £50. How much would the same kind of carpet cost for a passage 3 m by 1 m?

8 A toy car is made to a scale of 1 : 25. What is the length and breadth of the model if the actual car is 5 m long and 2 m broad?

9 A boy is making a plan of a hall 16 m by 10 m, using a scale of 1 cm to 1 m. What are the dimensions of the hall on his plan, and what is its area on his plan?

10 The R.F. of a map is 1/500000.
   a If two towns are 80 km apart, how far apart are they on the map?
   b If two towns are 2·4 cm apart on the map, what is the actual distance between them?

Exercise 8B

1 When going to Belgium on holiday a man changed £45 to francs at the rate of 120 francs to the £. After spending 3900 francs how many francs had he left?

2 A book originally contained 240 pages, with an average of 300 words per page. When it was reprinted with smaller type, 360 words per page were set. How many pages were needed?

3 A record played at 45 rpm takes 13 minutes. If a boy played it at $33\frac{1}{3}$ rpm, how long would it take?

4 A photographic slide measures 15 mm by 12 mm. When projected on to a screen, the height of the picture is 1 metre. By considering the ratios of the length and breadth of the slide and of the picture to be equal, calculate the breadth of the picture (which is greater than the height).

5 A fitted carpet for a room 4 m by 3·5 m costs £72. How much would the same kind of carpet cost for a room 3·5 m by 3 m?

6 The R.F. of a map is 1 : 200000.

Miscellaneous questions

   a If two towns are 64 km apart, how far apart are they on the map?
   b If two towns are 4·3 cm apart on the map, what is the actual distance between them?

7   Two cubes have edges 1·5 cm and 1 cm long. Find the ratios of:
   a the lengths of their edges   *b* their surface areas
   c their volumes.

8   A builder contracts to complete a job in 48 days, and takes on 14 men to do so. After 16 days the work is held up by bad weather for 11 days. How many extra men would be needed now to finish the job on time?

9   The R.F. of a plan is 1/5000. What land area is represented by a rectangular part of the plan 27 cm by 20 cm, if 30% of the area represents water?

10   An aircraft flying at 960 km/h takes 2 hours 15 minutes for a journey. How long (to the nearest minute) would it take at 1040 km/h?

Arithmetic

Summary

                First variable   *Second variable*
                       $a \longleftrightarrow c$
                       $b \longleftrightarrow d$

1   If the ratios $\dfrac{a}{b}$ and $\dfrac{c}{d}$ are equal, or

if the ratios $\dfrac{a}{c}$ and $\dfrac{b}{d}$ are equal, then the variables are related by *direct proportion*.

    Corresponding variables increase or decrease in the *same ratio* (e.g. if one is doubled, the other is doubled).

    The graph showing the relationship between the variables consists of a set of points which may be joined by a straight line passing through the origin.

2   If $\dfrac{a}{b} = \dfrac{d}{c}$, i.e. $ac = bd$, the variables are related by *inverse proportion*.

    As one variable increases, the corresponding one decreases in the *inverse ratio* (e.g. if one is doubled, the other is halved).

3   The *scale of a map* is often given by the *representative fraction*, i.e.

$$\frac{\text{distance on the map}}{\text{corresponding distance on the ground}}.$$

Note to the Teacher on Chapter 3

This chapter aims to give an elementary introduction to the idea of probability in such a way that it will be possible at a later stage to build on these foundations and to develop the sophistication both of the methods and the mathematics.

The experimental approach in *Section* 1 is fundamental in that it allows intuitive notions to crystallize in the definition given for probability on page 186. Certain terms have been used, including 'random experiment', 'outcome', 'relative frequency', 'probability', 'equally likely', but the new language is kept to the minimum. The background of the chapter contains terms like:

> *random experiment or activity* (e.g. tossing a coin)
> *trial* (a performance of the experiment)
> *outcome* (result of the trial, e.g. head or tail)
> *sample space* (the set of all possible outcomes, e.g. {H, T})
> *event* (a subset of the sample space—an outcome, or set of outcomes).

Teachers may wish to introduce these terms at appropriate stages.

In *Sections* 1 and 2 it is stressed that we are dealing with random experiments (in the sense that there is an outcome for each trial, but that the exact outcome cannot be predicted) for which every outcome has a probability which can be estimated through trials and relative frequencies. Further, in some simple cases the exact probability can be determined (e.g. tossing a coin, rolling a die) on the basis of an 'equally likely' definition.

It is hoped that Figures 2 and 3 will give a striking visual indication of the fact that in large numbers of trials the relative frequencies tend to cluster about, or stabilize in, a number, or 'limit', which is defined as the probability of the outcome. It should be realized that this 'limit' is approached in the sense that as the number of trials increases it becomes more and more likely that the relative frequency is close to this limit.

Section 2 gives a theoretical definition of probability which will be readily accepted after the experience obtained in *Section* 1, and Exercise 4 applies this in a wide variety of situations.

Further use of the definition of probability is made in *Section* 3 where expected frequencies are calculated on the simple basis of 'probability of outcome × number of trials'.

The fact that $0 \leqslant P \leqslant 1$, is looked at briefly in *Section* 4, where mention is made of the result that if the probability of an outcome is P, then the probability that the outcome will not happen is $1-P$, which is often useful in shortening calculations.

Section 5 introduces two rather difficult ideas concerning mutually exclusive outcomes and independent outcomes, and it is suggested that not all pupils should take this work at this stage.

The approach is based on visual arrays in order to try to present the ideas without introducing an undue (but almost necessary) degree of rigour. Behind the first result, $P(A \text{ or } B) = P(A) + P(B)$ is the statement 'If $A \cap B = \emptyset$, then $P(A \cup B) = P(A) + P(B)$,' and in fact the second idea, of independence, is often *defined* as follows: 'If $P(A \cap B) = P(A).P(B)$, then events A and B are said to be *independent* of each other.'

In the text for *Section* 5(i), we could of course use two dice, one after the other, but colours are clearer in the table. While all the examples in Exercise 5B and 6B are based on arrays with a visual analysis of the various sets of outcomes, it is of course possible to treat them at greater depth, and some teachers may wish to do so.

Some teachers may also wish to introduce the idea of a 'mathematical model' of a situation and apply it to some of the work.

(facing page 185)

Introduction to Probability

1 Experiment

A game of football is about to start. The referee tosses a coin; the captain of the visiting team calls 'Heads' or 'Tails'. If he is correct, he chooses which way to play.

Do you think that this is a fair way to decide? Why? Exercise 1 will enable you to think further about your answer.

In this chapter we will carry out a number of *experiments*, and study their *outcomes*. These will be *random experiments* in that the exact outcome of each trial cannot be predicted. For example:

(i) *Experiment:* Tossing a coin.
Outcomes: Heads, or Tails.
(ii) *Experiment:* Rolling a die.
Outcomes: 1, 2, 3, 4, 5 or 6 turns up.
(iii) *Experiment:* Choosing a card from the 'Hearts' in a pack.
Outcomes: Ace, king, queen, jack, 10, 9, 8, 7, 6, 5, 4, 3 or 2 of Hearts.

| Arithmetic | 186 |

Exercise 1

Experiment 1: Tossing a coin.
Outcomes: Heads, or Tails.

a Copy this table into your notebook.

Number of trials	25	50	75	100
Number of Heads				
Relative frequency of Heads = $\dfrac{\text{number of Heads}}{\text{number of trials}}$				

b Toss the same coin 25 times, and ask a partner to record the number of Heads (using tally marks, grouped in fives).

Have another 25 tosses, and enter the number of Heads for the 50 tosses in the table.

Continue until you have made 75 trials, then 100 trials.

c Calculate the *relative frequency* of Heads for 25, 50, 75 and 100 trials, rounding off your answers to 2 decimal places.

d Illustrate the relative frequencies as shown in Figure 2.

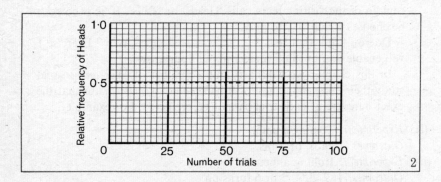

What do you notice about the relative frequencies?

e Combine all the trials in the class to calculate:

$$\text{Relative frequency of Heads} = \frac{\text{number of trials giving Heads}}{\text{total number of trials}}$$

As the number of trials increases the relative frequencies tend to cluster round a certain number, which is called the *probability* of the outcome.

Experiment

The probability of an outcome in a random experiment is the 'limit' of the relative frequencies of the outcome in large numbers of trials.

What do you estimate the probability of a Head to be?

Experiment 2: Rolling a die.
Outcomes: 1, 2, 3, 4, 5 or 6 turns up.

a Copy this table into your notebook.

Number of trials	6 36 60 120 180 240 300
Number of times 6 turns up	
Relative frequency of '6 turns up' = $\dfrac{\text{number of 6s}}{\text{number of trials}}$	

b As before, work in pairs. Toss a die 60 times, noting the number of 6s after 6, 36 and 60 trials.

Complete the number of 6s for 120, 180, 240 and 300 trials by combining your results with those of other pupils.

c Calculate the *relative frequency* of '6 turns up' for 6, 36, ..., 300 trials, to 2 decimal places in each case.

d Illustrate the relative frequencies as shown in Figure 3.

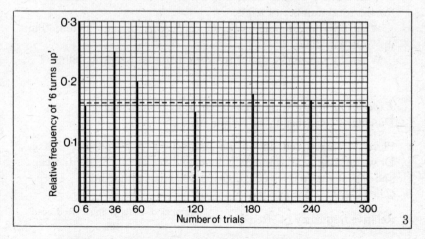

3

What do you estimate to be the *probability* of the outcome '6 turns up'?

e Express $\frac{1}{6}$ as a decimal fraction rounded off to 2 decimal places, and compare this number with your probability in *d*.

Arithmetic

f Do you think that rolling a die is a fair way to make a score in a game where two or more persons are playing? Why?

* * *

Experiment 3: Dropping a drawing pin on the floor.
Outcomes: 'Pin up', or 'Pin down', as in Figure 4.

4

a Drop 100 similar drawing pins all at once from a height of about 50 cm on to a level floor or table.
b One pupil should count the number 'Pin up', and another should count the number 'Pin down'. How will this check the answers?
c Calculate, to 2 decimal places:

$$\text{Relative frequency of 'Pin up'} = \frac{\text{number of trials with 'Pin up'}}{100}$$

$$\text{Relative frequency of 'Pin down,} = \frac{\text{number of trials with 'Pin down'}}{100}$$

d Repeat *a* and *b* several times and calculate the relative frequencies for the total number of trials.

What do you estimate to be the *probability* of the outcome 'Pin up'? Of 'Pin down'?

What do you notice about the sum of these two probabilities?

* * *

Experiment 4: Drawing a coloured bead from a bag.
Outcomes: 'Red bead', or 'White bead'.

a Place 50 mixed red and white beads in a non-transparent bag. Draw out one bead, note its colour, and replace it. Shake the bag, and repeat the trial 99 times, giving 100 trials altogether.
b Calculate, to 2 decimal places:

$$\text{Relative frequency of 'Red bead'} = \frac{\text{number of trials giving a 'Red bead'}}{100}$$

c What is your estimate of the *probability* of the outcome 'Red bead'.
d Compare your answer to *c* with the fraction of the number of red beads to the total number of beads in the bag.

* * *

Theory

Experiment 5: Spin a spinner like one of those in Figure 5.
Outcomes: 1, 2, 3, or 4 is scored.

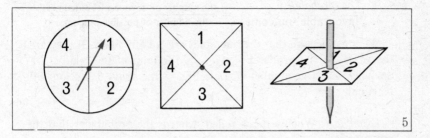

Suggest and carry out an experiment to estimate the probability of scoring 3 with the spinner.

2 Theory

a In the first experiment of Exercise 1 (Tossing a coin) we could argue as follows:

A coin is almost symmetrical (why 'almost'?). If it is tossed and allowed to land on a hard level surface, it is most unlikely that it will stand on its edge. Because of its symmetry, we see no reason for it to turn up 'Heads' more frequently or less frequently than 'Tails'. It follows that in a large number of trials we would expect to get equal numbers of Heads and Tails, so that the fraction of Heads in the total number of trials would be $\frac{1}{2}$.

b Repeat the above argument for the case of rolling a six-sided die.

* * *

When we see no reason why one outcome of an experiment should occur more frequently than another, we say that the outcomes are *equally likely*.

Thus a 'Head' and a 'Tail' are equally likely to turn up on the toss of an unbiased coin, and each of the six numbers on a die is equally likely to turn up when the die is rolled.

Arithmetic

In such cases, we say that where an experiment has several equally likely outcomes:

$$\text{The } \textit{probability} \text{ of a 'favourable' outcome} = \frac{\text{the number of 'favourable' outcomes}}{\text{the number of possible outcomes}}$$

We must be sure that the outcomes *are* equally likely. If they are, the values of the probability given by the limit of the relative frequencies in a large number of trials, and by the above definition, will be equal.

Example 1. When a die is rolled, what is the probability that the outcome is:

a '5 turns up'   *b* 'An even number turns up'?

In both cases the set of possible outcomes is {1, 2, 3, 4, 5, 6}, containing 6 members.

In *a* the subset of favourable outcomes is {5}, with one member. So the probability that '5 turns up' is

$$\frac{\text{number of 'favourable' outcomes}}{\text{number of possible outcomes}} = \frac{1}{6}$$

In *b* the subset of favourable outcomes is {2, 4, 6}, with three members. So the probability that 'An even number turns up' is $\frac{3}{6}$, or $\frac{1}{2}$.

Example 2. A bag contains 10 red beads and 30 black beads.

a What is the probability that when a bead is drawn from the bag it will be red?

b If it is red, and another is then drawn, what is the probability that it will be red also?

Briefly, writing P for probability, we have:

a P(red bead) = $\frac{10}{40} = \frac{1}{4}$

b P(red bead) = $\frac{9}{39} = \frac{3}{13}$

Exercise 2A

1. A game of chance consists of spinning an arrow which is equally likely to come to rest pointing to any one of the numbers 1, 2, 3, 4, 5 as shown in Figure 6. What is the probability that it will point to:

a 2   *b* an even number   *c* 5   *d* an odd number?

Theory

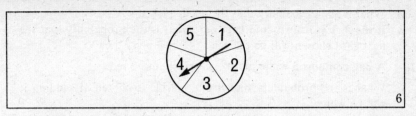

6

2 A box contains 10 white beads and 20 black beads. If a bead is drawn, what is the probability that it is:

a white   *b* black?

What is the sum of the two probabilities?

3 The names of the 25 boys and 10 girls in a class are written on similar pieces of paper, and placed in a hat. If one is drawn out at random, what is the probability that it is the name of:

a a boy   *b* a girl?

What is the sum of the two probabilities?

4 If a whole number from 1 to 20 inclusive is selected, and if each number has an equal chance of being selected, what is the probability that the number will be:

a even   *b* greater than 15   *c* prime?

5 A letter is chosen at random from the letters of the word ORANGE. What is the probability that it is:

a N   *b* a vowel?

6 When a card is drawn at random from a pack of 52 playing cards, what is the probability that it is:

a an ace   *b* the ace of spades   *c* a heart
d a king, queen or jack?

7 What is the probability that on rolling a die the number that turns up will be:

a odd   *b* less than 3   *c* 6   *d* prime?

8 There are 50 cars of the same make in a showroom. 18 are blue, 15 white, 10 green, 7 red. If it is equally likely that any one of them will be the first to be sold, what is the probability that it will be:

a white   *b* green   *c* red or blue   *d* neither red nor blue?

9 Five cards—ten, jack, queen, king, ace—are shuffled face downwards. One of the cards is picked at random.

| Arithmetic |

   a   What is the probability that the ace is picked?
   b   If the jack is drawn, and kept out, what is the probability that the next card chosen will be the ace?

10 A bag contains 5 white marbles, 3 black and 2 red.

   a   What is the probability that if one marble is chosen at random it will be white?
   b   If in fact a white marble is chosen, and not replaced, what is the probability that a black marble will be chosen next?

11 If you ask your friends to state any number from 1 to 10 inclusive, do you think that all the numbers are equally likely to be given? Is there such a thing as a lucky number?

Exercise 2B

1 If a letter is chosen at random from the letters of the word MISSISSIPPI, what is the probability that it is:

   a S   *b* I   *c* neither I nor S?

2 A symmetrical triangular pyramid (tetrahedron) has the numbers 1, 2, 3, 4 painted on its faces. Find the probability that when it is tossed it will land so that:

   a the 2-face is downwards
   b the sum of the numbers on the three faces showing is even.

3 If there are 500000000 Premium Bonds eligible for the draw each week and you own 250 Bonds, what is the probability that you will win the single £25000 prize in any one week?

4 In a class of 30 pupils, 18 like pop records only, 7 like classical records, the rest like neither. If a pupil in the class is chosen at random, what is the probability that he likes:

   a pop records   *b* neither kind of record?

5 If all the aces, kings, queens and jacks are removed from a pack of cards, what is the probability that when a card is selected at random from the remainder it will be:

   a a 7   *b* a king   *c* greater than 7
   d a card with a prime number?

6 Figure 7 shows the distribution of boys (B) and girls (G) at the desks of a classroom. If the name of each pupil is written on a piece

Theory

of paper and placed in a box, state the probability that when a paper is drawn at random from the box it will contain the name of:

a a girl *b* a boy *c* a pupil sitting on his or her own
d a pupil sitting next to another pupil
e a pupil sitting next to a wall
f a pupil sitting at the back of the room
g a pupil sitting directly behind a girl
h a pupil sitting directly in front of a boy.

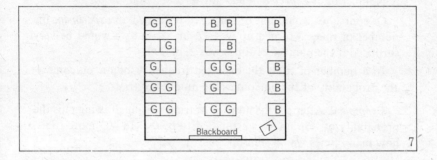

7 A box contains three white and two black counters. What is the probability that when a counter is drawn at random it will be white?
   If the counter is white, and is not replaced, what is the probability that the next one will be white also?

8 A carton of twelve pens contains three faulty pens. What is the probability that a pen chosen at random will be faulty? If it is not faulty, and is not replaced, what is the probability that the next pen chosen will not be faulty?

9 A table-tennis ball is dropped from a height on to a 10×10 grid of square holes large enough to contain the ball. Why is the probability of the ball landing in any particular hole not $\frac{1}{100}$? Discuss this.

Arithmetic

3 Calculating expected frequency

Example 1. If we roll a die 300 times how many times will we expect 3 to turn up?

$$P(3) = \tfrac{1}{6}.$$

Number of 3s *expected* in 300 throws $= \tfrac{1}{6} \times 300 = 50$.

Of course we would not be surprised if in such an *experiment* the number of times 3 turned up was 47, or 52; but we would be very surprised if the number of times was 2, or 290.

In a number of trials the expected frequency of an outcome = the probability of the outcome × the number of trials.

Example 2. After a large number of trials tossing drawing pins the probability of 'Pin up' was estimated to be 0·3. In 400 more trials, how many times would 'Pin up' be expected?

$$P(\text{'Pin up'}) = 0 \cdot 3$$

Number of times 'Pin up' is expected in 400 trials $= 0 \cdot 3 \times 400$
$= 120.$

Exercise 3

1 If a coin is tossed 100 times, what is the expected frequency of 'Heads'?

2 If a die is rolled 60 times, what is the expected frequency of:
  a 1 turns up *b* a number divisible by 3 turns up
  c a factor of 6 turns up?

3 If the arrow in Figure 6 (page 191) is spun 100 times, what is the expected frequency of:
  a a 5 *b* an even number?
  If the arrow is spun 1000 times, what final score would you expect if all the individual scores are added together?

4 It was found that the probability of a child getting measles was 0·13. Out of 1200 children how many would you expect to catch measles?

5 A pack of cards is shuffled and then one card is chosen at random. The card is replaced, and the experiment is made 1000 times

Probability of certain success and certain failure

altogether. How many times would you expect an ace to be chosen? How many times the ace of spades?

6   An insurance company estimates that the probability of a driver having an accident in any year is 0·14. Out of 250 drivers, how many can be expected to have an accident this year?

7   Over a great many attempts it is found that the probability of a marksman scoring a 'bull's-eye' is 0·92. How many bull's-eyes would you expect him to score in a competition where he fires 80 shots?

    Is it possible to calculate the probability 0·92 for the marksman?

8 *a* List the set of four possible outcomes when two coins are tossed.
  b How often would you expect to obtain two heads in 200 trials?
  c Would you be surprised to obtain two heads
    (*1*) 53 times   (*2*) 185 times   (*3*) not at all?

4 Probability of certain success and certain failure

Exercise 4

1   A two-headed coin is tossed. What is the probability that:
 a a head turns up   *b* a tail turns up?

2   A certain die has the number 6 on *each* face. What is the probability on rolling the die that:
 a 6 turns up   *b* 3 turns up   *c* an even number turns up
 d an odd number turns up?

3 *a* What is the probability of an outcome which is certain to happen?
  b What is the probability of an outcome which cannot take place?
  c Between what two numbers must the probability of an outcome lie?

Probabilities can be illustrated on a number line as in Figure 8.

```
Probability... 0              0·5              1
Outcome... Impossible     As likely as not    Certain
```

8

Arithmetic

4 Sketch the line in Figure 8 and place the probabilities of the following on the line:
  a that the sun will rise tomorrow
  b that when I toss a penny a Head or Tail will show
  c that when I roll a die a 7 will turn up
  d that a month chosen at random from the calendar will be June
  e that 1980 will be a leap year
  f that some day I'll die
  g that a day in the week chosen at random will be Saturday or Sunday.

5 For the toss of a coin, calculate:
  a P(Head)   *b* P(Tail)   *c* P(Head) + P(Tail)

6 For the roll of a die, calculate:
  a P(even number)   *b* P(odd number)
  c P(even number) + P(odd number)

7 For the random selection of a bead from a bag containing 15 blue beads and 12 red beads, calculate:
  a P(blue bead)   *b* P(red bead)
  c P(blue bead) + P(red bead)

8 Copy and complete this statement for an outcome A of a given experiment.
P(outcome A) + P(outcome not-A) = ...

> If the probability of an outcome of an experiment is P, then the probability that the outcome will not happen is $1 - P$.

9 The probability of scoring 12 when throwing two dice at once is $\frac{1}{36}$. What is the probability of the score not being 12?

10 If the probability that it will rain tomorrow is 0·35, what is the probability that it will not rain tomorrow?

11 A spinner is equally likely to point to any one of the numbers 1, 2, 3, 4, 5, 6, 7. What is the probability of scoring a number divisible by 3?

Calculate in two different ways, the probability of scoring a number not divisible by 3.

5 Combining outcomes

(i) *Mutually exclusive outcomes*

Suppose that a blue die and a black die are rolled. The set of all possible outcomes is shown in the table below.

		\multicolumn{6}{c}{Black die}					
		1	2	3	4	5	6
Blue die	1	(1,1)	(1,2)	(1,3)	(1,4)	(1,5)	(1,6)
	2	(2,1)	(2,2)	(2,3)	(2,4)	(2,5)	(2,6)
	3	(3,1)	(3,2)	(3,3)	(3,4)	(3,5)	(3,6)
	4	(4,1)	(4,2)	(4,3)	(4,4)	(4,5)	(4,6)
	5	(5,1)	(5,2)	(5,3)	(5,4)	(5,5)	(5,6)
	6	(6,1)	(6,2)	(6,3)	(6,4)	(6,5)	(6,6)

From the table a total score of 5 can be obtained in 4 ways (4, 1), (3, 2), etc. Hence the probability of getting a total score of 5 is $\frac{4}{36}$, i.e. $\qquad P(5) = \frac{4}{36} = \frac{1}{9}$; and $P(10) = \frac{3}{36} = \frac{1}{12}$

Also from the table, $P(5 \text{ or } 10) = \frac{7}{36}$

Notice that $\qquad P(5 \text{ or } 10) = P(5) + P(10)$

This form of result can be used in general when the sets of outcomes are quite separate, that is when there is no member which appears in both sets. In this case we say that the outcomes are *mutually exclusive*.

If A and B are mutually exclusive outcomes,

then $\qquad P(A \text{ or } B) = P(A) + P(B)$

Exercise 5B

1 a Copy the table for the rolling of two dice, and write down the following probabilities for the total scores on the two dice:

$$P(2), P(3), P(4), ..., P(12).$$

b Are all of these outcomes equally likely?

2 What is the most likely, and the least likely, score on rolling two dice?

Arithmetic

3 What is P(2 or 12) for the two dice?

4 What is P(3 or 4 or 5) for the two dice?

5 What is P(prime number) for the two dice?

6 What is P(score greater than 7) for the two dice?

7 What is the probability of getting a score which:

  a includes a 1 on the blue die

  b includes a 1 on the blue die or a 6 on the blue die?

8 A spinner is equally likely to point to 1, 2 or 3. It is spun twice. Make a table of ordered pairs (First spin, Second spin). Find the following from the table, and also by calculation:

  a P(a double, i.e. (1, 1), (2, 2), etc.) **b** P(total score of 3 or 5)

  c P(1 with first spin or 3 with first spin)

9 When two dice are rolled, what is the probability of an outcome in which the score on the second die is greater than that on the first?

(ii) *Independent outcomes*

Again a blue die and a black die are rolled, and the possible outcomes are shown in the table below.

	Black die					
	1	2	3	4	5	6
1	(1,1)	(1,2)	(1,3)	(1,4)	(1,5)	(1,6)
2	(2,1)	(2,2)	(2,3)	(2,4)	(2,5)	(2,6)
Blue 3	(3,1)	(3,2)	(3,3)	(3,4)	(3,5)	(3,6)
die 4	(4,1)	(4,2)	(4,3)	(4,4)	(4,5)	(4,6)
5	(5,1)	(5,2)	(5,3)	(5,4)	(5,5)	(5,6)
6	(6,1)	(6,2)	(6,3)	(6,4)	(6,5)	(6,6)

From the table, P(blue 4) = $\frac{6}{36}$ = $\frac{1}{6}$; the set of favourable outcomes is shown in the horizontal box.

Also, P(black 4) = $\frac{6}{36}$ = $\frac{1}{6}$; the set of favourable outcomes is shown in the vertical box. And P(blue 4 *and* black 4) = $\frac{1}{36}$; the set of favourable outcomes is shown in the *intersection* of the two boxes.

Notice that P(blue 4 *and* black 4) = P(blue 4) × P(black 4).

This result is generally true when the two outcomes occur *inde-*

Combining outcomes

pendently of each other (in this case the blue 4 and the black 4 occur quite independently of each other).

If A and B are outcomes which are independent of each other,
$$P(A \text{ and } B) = P(A) \times P(B)$$

Exercise 6B

1. With the assistance of the table of outcomes of rolling two dice, find P(double 6). Now *calculate* P(double 6).

2. Find P(blue 2 and black 5) from the table, and by calculation.

3. Find P(blue 1 and black number greater than 3) from the table, and by calculation.

4. Copy and complete this table for the toss of a coin and the roll of a die:

		Die					
		1	2	3	4	5	6
Coin	H	(H, 1)	(H, 2)	.	.	.	.
	T	.	.	.	.	.	(T, 6)

    a. How many members are there in the set of possible outcomes?
    b. Show by a box the subset of outcomes containing a Tail.
    c. Show by a box the subset of outcomes containing 4.
    d. From the intersection of these subsets, what is P(Tail, 4)?
    e. From the table, what is P(Tail), and what is P(4)?
    f. Verify that P(Tail, 4) = P(Tail) × P(4).

5. Repeat question *4e* and *f* for P(Head), P(even number) and P(Head, even number).

6. Make a table for the toss of two coins, putting 'First Coin' on the left, 'Second Coin' at the top.
   Find P(H, H), P(T, T) and P(a head and a tail, in any order).

7. A spinner is equally likely to point to any one of 1, 2, 3, 4. Make a table of ordered pairs (First spin, Second spin). Find the probability of:
    a. two even numbers
    b. two odd numbers
    c. an even number followed by an odd number
    d. an odd number followed by an even number.
    What is the sum of the four probabilities? Why?

8. The spinner as in question *7* is spun once, then a die is rolled. Make

up a table of ordered pairs (Spinner, Die). Hence find (where E means Even and O means Odd).

a P(E, E)   *b* P[(E, O) or (O, E)]   *c* P(total of 10)
d P(total of 1)   *e* P(total less than 6)

9 Copy and complete this array of ordered triples for the possible outcomes when 3 coins are tossed simultaneously:

HHH	HHT	HTH	HTT
THH			

Hence find the probability of getting:

a 3 Heads   *b* 2 Heads and a Tail in any order.   *c* 3 Tails

What is the probability of getting 1 Head on tossing 1 coin? 2 Heads from 2 coins? 3 Heads from 3 coins? 4 Heads from 4 coins? 5 Heads from 5 coins?

10 Two travellers visit the same shop in the same week (Monday to Friday). Each is equally likely to visit the shop on any day. What is the probability that they will visit the shop:

a on the same day   *b* on consecutive days?

Summary

1. The *probability* of an outcome in a random experiment is the limit of the relative frequencies of the outcome in large numbers of trials.

    Where an experiment has several *equally likely* outcomes, the probability of a 'favourable outcome'
    $$= \frac{\text{the number of 'favourable' outcomes}}{\text{the number of possible outcomes}}$$

2. If the probability of an outcome of an experiment is P, then the probability that the outcome will not happen is $1-P$.

3. $0 \leq P \leq 1$.

4. In a number of trials the expected frequency of an outcome = the probability of the outcome × the number of trials.

5. For two mutually exclusive outcomes A and B of an event,
    $$P(A \text{ or } B) = P(A) + P(B)$$

6. For two independent outcomes A and B of an event,
    $$P(A \text{ and } B) = P(A) \times P(B)$$

Arithmetic

Topics to explore

1   Find out the number of pupils in your class who have birthdays on the same date. The *probability* of this is greater than you think. For a group of 20 persons it is 0·41, for 25 it is 0·57, for 30 it is 0·71, for 40 it is 0·89.

2   Are all the figures 0, 1, 2, ..., 9 equally likely in a telephone number? Investigate this for the last figure of all the numbers in a page of the telephone directory.

     Repeat the investigation for the first figure of all the numbers in a page.

Note to the Teacher on Chapter 4

With the changeover from the British to the metric system in money, weights and measures, many traditional methods in calculation are now discarded. However there seems no likelihood in the foreseeable future that the table of time—days, hours, minutes, seconds—will be altered. It is important, therefore, that the special methods necessary to deal with this table are not overlooked. *Section* 1 is concerned with a topic of everyday interest and affords an opportunity to practise *addition* and *subtraction* in these units. The use of the 24-hour clock is almost universal in timetables, but there is a lack of uniformity in the conventions for writing times; Airways appear to favour 0530, British Railways 05 30 and Buses 05.30. Experience has shown that pupils are confused by the first method—1000 is often read as one thousand hours; the third method leads to the error of reading the time as a decimal, or assuming that there are 100 minutes in an hour; so it has been decided to adopt the second method. The teacher should stress that midnight can be shown either as 00 00 or as 2400; and it may be pointed out that travel organizations try to avoid starting a train or plane at midnight, preferring to start at 23 59 which avoids misunderstandings about the day of the week.

Pupils have studied proportion, and *Section* 2 stresses that a graph illustrating direct proportion is a straight line. However, this section is dealing with average speeds, and it should be pointed out that neither of the graphs in Figure 2 gives an exact picture of what is happening. An attempt is made to show in a very simple way that the slope (gradient) of the distance–time graph gives an indication of the speed.

The relationship between Distance, Speed and Time is of great importance in modern life. Whichever of the methods *a* and *b* of *Section* 3 is adopted (the teacher can make a choice) it is necessary to be able to *multiply* and *divide* in the system of time units. In multiplication it is suggested in Example 1 of *Section* 3 that the traditional method (i.e. $56 \times 1\frac{1}{6} = 56 \times \frac{7}{6} = \frac{392}{6} =$ etc.) might be replaced as follows. To calculate the distance covered in 4 h 15 min by a plane travelling at 675 km/h:

$$4 \times 675 \text{ km} = 2700 \text{ km}$$
$$\tfrac{1}{4} \times 675 \text{ km} = \underline{168 \cdot 7}$$
$$4\tfrac{1}{4} \times 675 \text{ km} = 2869 \text{ km}$$

(Compare this with the recommended method in Book 1 Chapter 3 and Book 3 Chapter 1 for calculating a percentage of a quantity.) In division the two methods used in the course are:

(i) $\dfrac{28}{5\tfrac{1}{3}} = 28 \div \dfrac{16}{3} = 28 \times \dfrac{3}{16}$ (*multiplying* by the reciprocal), etc.

(ii) $\dfrac{28}{5\tfrac{1}{3}} = \dfrac{28 \times 3}{5\tfrac{1}{3} \times 3} = \dfrac{28 \times 3}{16}$, etc.

Pupils must be proficient in the use of one or both methods.

Exercises 1 and 2 should be within the capabilities of most pupils, but questions *12* to *16* of Exercise 3 could be omitted by less able pupils. Throughout, answers should be given to a reasonable degree of approximation where appropriate, often rounded to 3 significant figures (e.g. Exercise 3, question *2e*).

(facing page 203)

4 Time, Distance, Speed

1 Timetables

Time of day in a timetable is usually given in terms of the 24-hour clock. The time, measured from midnight, is written 0530 or 05 30 or 05.30. The first two figures give the *hour*, the last two figures give the *minutes* past that hour. Note that midnight is 0000 at the start of the day and also 2400 at the end of the day. Midday is 1200.

Example 1. 0530 means 5 hours 30 minutes after midnight, that is half past five in the morning (5.30 am).

Example 2. 1952 means 19 hours 52 minutes after midnight, that is 52 minutes past 7 pm or 8 minutes to 8 in the evening.

When subtracting two times, do not forget
(i) that there are 60 minutes in 1 hour
(ii) that when a journey starts one day and finishes the next day it is best to find the time *till* midnight (2400) the first day, and then to add the time *from* midnight (0000) the second day.

Example 3. A journey takes from 0535 till 1820 the same day. How long is this?

```
    1820      or    0535 till 0600 is      25 min
   −0535            0600 till 1820 is  12 h 20 min
   ─────                                ───────────
    1245            Total time is      12 h 45 min
```

Example 4. What time passes between 1935 one day and 0355 the next day?

```
    2400           0425
   −1935          +0355
   ─────          ─────
    0425           0820    (80 min = 1 h 20 min)
```

The time that passes is 8 hours 20 minutes.

Arithmetic

Exercise 1

Questions *1–6* refer to the following Bus Timetables.

Glasgow to Inverness				Inverness to Glasgow			
Glasgow	lve	0945	2350	Inverness	lve	1000	2300
Stirling	arr	1048	0040	Carrbridge		1048	2347
	lve	1048	0045	Kingussie	arr	1128	0026
Perth	arr	1215	0150		lve	1138	0026
	lve	1215	0220	Pitlochry	arr	1310	0155
Dunkeld	arr	1245	0250		lve	1405	0155
	lve	1335	0250	Dunkeld		1430	0220
Pitlochry		1400	0315	Perth	arr	1500	0250
Kingussie	arr	1532	0444		lve	1500	0320
	lve	1542	0444	Stirling	arr	1612	0425
Carrbridge		1622	0523		lve	1617	0430
Inverness	arr	1710	0610	Glasgow	arr	1720	0525

1. How long does it take from Pitlochry to Inverness on each of the two services?

2. What are the times of departure and arrival in am and pm notation at Glasgow and Inverness?

3. Where will you have time for lunch on each of the day-time services? How long in each case?

4. Where will you have time for a cup of tea on the night-time services? How long in each case?

5. What is the total time for each of the four services between Glasgow and Inverness?

6. Find the total time for which the bus is stationary during each of the four services. Hence, using your answers to question *5*, find the total *running* time in each case.

7. Car ferries leave Newhaven for Dieppe daily in July at 0415, 0730, 1135, 1545, 1900 and 2345. The crossing takes $3\frac{3}{4}$ hours. What is the scheduled time of arrival of each ferry?

Questions *8–10* refer to this timetable for the Sealink (train and boat).

Distance–time graphs

Summer service from London to Cologne via Ostend:

London	dep	0800	1100	1400	1500	2300
Folkestone	arr	0925	—	—	—	—
	dep	0955	—	—	—	—
Dover	arr	—	1220	1520	1620	0022
	dep	—	1300	1600	1700	0100
Ostend	arr	1335	1620	1920	2020	0430
	dep	1451	1657	2110	2115	0615
Brussels		1609	1817	2232	2237	0732
Aachen	arr	1758	2020	0042	0050	0845
	dep	1800	2034	0102	0110	0930
Cologne	arr	1842	2116	0144	0151	1024

8 Which service would you take if you had to start a coach tour at Ostend at 3 pm? if you had to meet a friend in Cologne at noon? if you wanted to reach Brussels as late as possible in the day?

9 Find the total time for each service from London to Cologne.

10 Find the time taken for the sea crossing (Folkestone or Dover to Ostend) on each service.

2 Distance–time graphs

A car-ferry sailing from Hull to Rotterdam takes 10 hours to sail the 250 km from the mouth of the River Humber to the River Maas. Clearly the average distance sailed per hour is $250 \div 10$ km, i.e. 25 km. We say that the average speed is 25 km per hour (25 km/h). We can imagine that if there were no changes in the wind or current the speed throughout the journey would remain constant; then in 1 hour the ship would travel 25 km, in 2 hours 50 km, and so on, as indicated by the dots in Figure 1, which give a graph of distance against time.

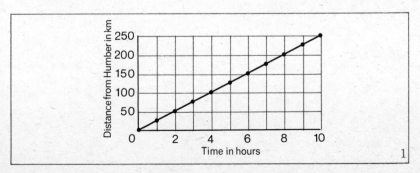

Arithmetic

Speed is rate of change of distance. When the speed is constant, the distance is proportional to the time and, as we saw in Chapter 2, the graph of a direct proportion relationship is a straight line.

In most journeys the speed is not constant. Think of a motorist travelling the 60 km from Manchester to Stoke-on-Trent. Suggest some road conditions which would cause changes in his speed. Look at this table:

Distance from Manchester in km	0	5	25	30	30	40	60
Time in hours	0	$\frac{1}{4}$	$\frac{1}{2}$	$\frac{3}{4}$	1	$1\frac{1}{4}$	$1\frac{1}{2}$

The average speed over the $1\frac{1}{2}$-hour period is

$$\frac{60}{1\frac{1}{2}} \text{ km/h} = \frac{60 \times 2}{1\frac{1}{2} \times 2} \text{ km/h} = \frac{120}{3} \text{ km/h} = 40 \text{ km/h}$$

But in the first $\frac{1}{4}$ hour he travels 5 km, so his average speed is 20 km/h; in the next $\frac{1}{4}$ hour he travels 20 km, so his average speed is 80 km/h. Continue to calculate his average speed for each $\frac{1}{4}$ hour. You can see that average speed must be interpreted very carefully.

Figure 2 shows the graph of distance against time for the given table. The black line shows the average speed in each section of the journey, while the broken line shows the average speed for the whole journey.

Can you see what happens to the *slope* of the graph as the motorist travels faster? slower? What can you say about the slope of the graph when the motorist stops for a cup of tea? From the graph, read off the $\frac{1}{4}$-hour periods during which the motorist is travelling *faster* than the average speed; *slower* than the average speed; *at* the average speed.

Distance–time graphs

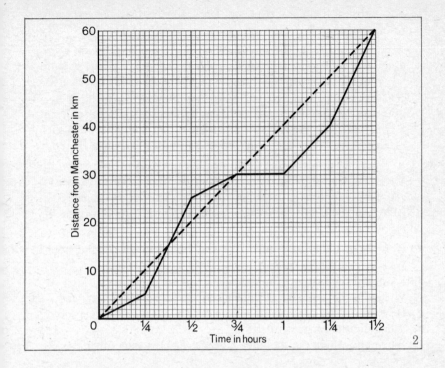

Exercise 2

In questions *1* and *2* draw the graphs on 5-mm squared paper, taking 5 squares horizontally to represent 1 hour and 5 squares vertically to represent 20 km.

1 On the same diagram draw distance–time graphs to show a 40-km journey

  a by a man walking at an average speed of 8 km/h. (On the graph paper show that in 1 hour he goes 8 km, in 2 hours 16 km, etc., and draw in a straight line.)
  b by a boy cycling at 20 km/h
  c by a lorry travelling at 40 km/h
  d by a motorist driving at 80 km/h.

2 Draw a graph to show the progress of two boys who set off at the same time from Luton to cycle to Reading, 80 km away, one at 24 km/h and the other at 16 km/h. From the graph find when each

Arithmetic

boy reaches Reading. How far ahead is the first boy when he reaches Reading?

3 A boy sets off at 0900 from Glasgow to cycle to Ayr, 50 km away, at an average speed of 12 km/h. After cycling for 2 hours he stops for $\frac{1}{2}$ hour for a picnic, then goes on as before. At 1100 his father sets off from Glasgow to Ayr in his car travelling at 40 km/h. Draw a graph on 2-mm squared paper, taking 2 cm horizontally to represent 1 hour and 2 cm vertically to represent 10 km, and find when each reaches Ayr. When and where does the father overtake the boy?

4 In Figure 3 the black line illustrates the progress of a lorry which leaves Perth at 0800 and travels towards Aberdeen, a distance of 130 km. The broken line illustrates the journey of a car from Aberdeen to Perth on the same road. From the graph answer the following:

a How far did the lorry go in the first hour? in the second hour? What is its average speed?
b When did the lorry reach Aberdeen?
c When did the car leave Aberdeen?
d How far did the car go in the first hour? What is its average speed?
e When did the car reach Perth?
f When and where did the lorry and the car pass each other?

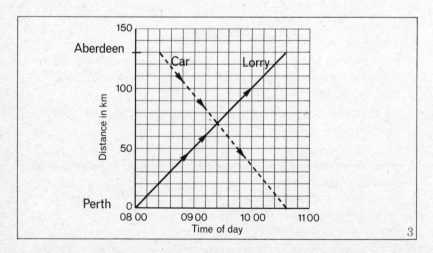

5 A BEA plane leaves Athens daily at 0900 and arrives in London, 2400 km away, at 1224. On 5-mm squared paper take 5 squares

Time–distance–speed calculations

horizontally to represent 1 hour (i.e. 1 square to represent 12 min) and 5 squares vertically to represent 500 km from Athens. By plotting the start and finish of the journey draw the graph. By finding the distance from Athens at 1000, find the average speed of the plane.

Another BEA plane leaves London daily at 0955, arriving in Athens at 1315. By drawing its graph on the same diagram, find when the planes pass one another, and at what distance from Athens. (Assume they follow the same flight path.) All times are British times.

3 Time–distance–speed calculations

Calculations on speed may be done in two ways.

(i) Proportion method

Example 1. A train runs from Glasgow to Edinburgh in 1 hour 10 minutes at an average speed of 56 km/h. What is the distance between the towns?

Time		Distance	
1 h	⟷	56 km	$56 \times 1 = 56$
$1\frac{1}{6}$ h	⟷	$56 \text{ km} \times 1\frac{1}{6}$	$56 \times \frac{1}{6} = \underline{9\cdot3}$
		$= 65$ km	$\underline{65\cdot3}$

Example 2. The distance between Ostend and Innsbruck by autobahn is 975 km. How long will the journey take at an average speed of 65 km/h?

Distance		Time
65 km	⟷	1 h
975 km	⟷	$1 \text{ h} \times \frac{975}{65}$
		$= 15$ h

(ii) Formula method

What distance is covered at 30 km/h in 5 hours?
What distance is covered at $12\frac{1}{2}$ km/h in 3 hours?
What distance is covered at S km/h in T hours?

Arithmetic

From these examples it is seen that, if D represents distance, S represents a constant speed and T represents time.

$$D = ST$$

Great care must be taken to check that the units of distance and time are the same as those of the speed.

Example 1. (As above)

$$\text{Speed} = 56 \text{ km/h} \qquad D = ST$$
$$\text{Time} = 1\tfrac{1}{6} \text{ h} \qquad \text{Distance} = 56 \times 1\tfrac{1}{6} \text{ km}$$
$$= 65 \text{ km}$$

Example 2. (As above)

$$\text{Distance} = 975 \text{ km} \qquad D = ST$$
$$\text{Speed} = 65 \text{ km/h} \qquad 975 = 65T$$
$$\Leftrightarrow T = \tfrac{975}{65}$$
$$\text{Time} = 15 \text{ hours}$$

Exercise 3

1 In each of the following find the average speed in an appropriate unit (km/h, cm/s, etc.):

	a	b	c	d	e
Distance	150 km	440 m	980 km	54 cm	500 m
Time	4 h	55 s	14 h	4·5 s	40 min

2 In each of the following find the distance travelled:

	a	b	c	d	e
Average speed	48 km/h	600 km/h	$3\tfrac{1}{3}$ km/h	45 cm/s	300 m/s
Time	4 h	6 h 40 min	$4\tfrac{1}{2}$ h	$2\tfrac{1}{2}$ s	2 min

3 In each of the following find the time taken:

	a	b	c	d	e
Distance	4200 km	27 cm	440 m	1·7 m	3 km
Average speed	600 km/h	54 cm/s	55 m/s	85 cm/s	7·5 m/min

4 What average speed must I keep up to travel to Salzburg from Rotterdam, a distance of 1035 km, in 9 hours?

5 The Royal Scot train leaves Glasgow at 1000 and reaches London at 1645. If the distance is 624 km, what is the average speed, rounded off to the nearest km/h?

Time–distance–speed calculations

6 A space-craft travelling at an average speed of 17 600 km/h takes 21 h 30 min to reach the moon from launching time. Estimate the distance from the earth to the moon, to 3 significant figures.

7 An artificial satellite orbits the earth once every 2 h 10 min. If its average speed is 23 200 km/h, what is the length of its orbit, to the nearest hundred km.

8 A jet airliner leaves New York and flies at an average speed of 825 km/h. How long will it take to reach London which is 6600 km away?

9 If a man walks at an average speed of 6 km/h, how far will he walk in 25 minutes?

10 If a motor-car covers 30 km in 40 minutes find its average speed in km/h.

How long will a journey of $112\frac{1}{2}$ km take if the driver keeps up this average speed, but stops for half an hour for coffee on the way?

11 A boy can walk to school in 15 minutes. If he can cycle three times as fast as he can walk, how long does it take him to cycle to school?

The distance from his house to the school is 1200 m. Calculate his average walking speed and his average cycling speed in km/h.

12 The times of some trains between Aberdeen and Edinburgh are as follows:

Aberdeen	dep	0730	1205	1745	1915
Dundee	arr	0902	1342	1922	2055
	dep	0905	1352	1929	2100
Kirkcaldy	dep	—	1500	2025	2158
Edinburgh	arr	1030	1540	2105	2239

a If the trains which stop at Kirkcaldy halt there for 2 minutes, find the actual travelling time of each train between Aberdeen and Edinburgh.

b The distance between Aberdeen and Edinburgh is 190 km. Find the average speed of each train for the whole journey, to the nearest km/h, based on the *total* times for the journeys.

c On 2-mm squared paper draw graphs to illustrate the journeys of these trains, given that the distance from Aberdeen to Dundee is 105 km and from Dundee to Kirkcaldy is 45 km. Take a scale of 2 cm to 30 minutes horizontally and 2 cm to 50 km vertically.

13 The world record for the 100-metre race set in 1960 was 10 seconds. What average speed does this represent in km/h?

Arithmetic

Try to find the information necessary to work out the next three questions.

14. What is the speed of light? What is the distance between the sun and the earth? Use these facts to find how long it takes for light to travel from the sun to the earth.

15. Find the distance that a ray of light can travel in a year. (This is one of the astronomical units of distance the *light year*.)

16. What is the circumference of the earth at the equator? What is the speed of radio waves?

    Find the time taken by a radio wave to travel once round the earth at the equator.

Summary

1. A *time* given as 09 53 means
   9.53 am, or 7 minutes to 10 in the morning.

   A *time* given as 16 30 means
   4.30 pm, or half past 4 in the afternoon.

2. $Average\ Speed = \dfrac{Distance}{Time}$

3. If D represents distance, S represents speed and T represents time,
$$D = ST$$

Revision Exercises

Revision Exercises on Chapter 1
Social Arithmetic—1

Revision Exercise 1A

1. Calculate the total charge in a telephone bill which lists a rental charge of £5·25, 450 units at 1·2p per unit, and calls via the operator costing £2·19.

2. Calculate the cost of electricity at 3p per unit for the first 60 units, then 0·6p per unit for the remainder, between meter readings of 19 146 and 23 456.

3. Calculate the amount to be paid for gas if the standing charge is £3·75, and 270 therms of gas at 9·5p per therm have been used.

4. Make a ready reckoner which shows the cost of 1, 2, 3, ..., 10 litres of paraffin, given that 1 litre costs 3·5p. Use it to write down the cost of:
    - a 6, 60, 600 litres
    - b 3, 30, 300 litres
    - c 15 litres
    - d 34 litres
    - e 55 litres

5. At a sale, 20% discount is given on purchases; how much will you actually pay for:
    - a a coat marked £18
    - b a suit marked £25·20
    - c a hat marked £2·55
    - d 3 shirts marked £1·84 each?

6. Work out the following bill:

    9 rolls wallpaper   at 73p per roll
    3 rolls ceiling paper   at 42p per roll
    4 cans paint   at 57½p per can

    Allow 5% discount on the total for cash payment (ignore fractions of a penny).

Revision Exercises on Chapter 1

7. Calculate the interest in the following cases:
   a. Principal = £250, rate = 6% p.a., time = 1 year
   b. Principal = £82, rate = $3\frac{1}{2}$% p.a., time = 1 year
   c. Principal = £450, rate = 5% p.a., time = 8 months

8. Calculate the amount in the following cases:
   a. Principal = £10 000, rate = $6\frac{1}{2}$% p.a., time = 1 year
   b. Principal = £120, rate = 7% p.a., time = 6 months
   c. Principal = £92·13, rate = $2\frac{1}{2}$% p.a., time = 5 months

9. Calculate the rate per cent per annum:
   a. Principal = £250, interest = £20, time = 1 year
   b. Principal = £120, amount = £128, time = 1 year
   c. Principal = £36, interest = 75p, time = 4 months

10. Calculate the profit or loss per cent *as a percentage of the selling price* for each of the following:
    a. Cost price = £120, selling price = £144
    b. Cost price = 42p per dozen, selling price = 4p each
    c. Cost price = £1·44 for 36, selling price = 10p for 3

11. A dealer buys a grandfather clock for £48. What is his selling price if he makes:
    a. a profit of 100%
    b. a profit of $33\frac{1}{3}$%
    c. a loss of 15%
    d. a loss of 100%?

12. Express as percentages: a $\frac{3}{25}$ b $1\frac{1}{4}$ c 0·36 d 1·3

13. A football ground has seats for 5000 spectators in the stand and 25 000 spectators on the terracing.
    a. What percentage of the total accommodation is stand seats?
    b. If the stand is full and there are 4000 spectators on the terracing, what percentage of the total accommodation is occupied?
    c. If 30% of the stand seats are taken and 15% of the terracing space is occupied, how many spectators are at the match?

14. There are 118 800 spectators at a match at Hampden Park. If this fills 90% of the possible accommodation of the ground, how many can it hold when full?

| Arithmetic | 216 |

Revision Exercise 1B

1. Calculate the charge for electricity between meter readings of 12197 and 15153 if the cost of electricity is 3·2p per unit for the first 80 units and 0·5p per unit for the remainder.

2. The insurance payable on a man's car is £37·50. He gets a reduction of 60% as a no-claims bonus, and a further reduction of 5% on the net cost (after the no-claims bonus has been deducted) as commission. What does he actually pay?

3. When shopping at a store a customer gets one dividend stamp for every 5p spent. The stamps are stuck in a book which contains 30 pages, each page holding 40 stamps. How much must the customer spend to fill the book? The completed book can be exchanged for 40p. What percentage discount does this represent (to two significant figures).

4. Calculate the interest on £650 for 7 months at $8\frac{1}{2}\%$ p.a. to the nearest penny.

5. A man has £480 to invest. Which of the following will give him the better return, and by how much?
   - *a* To place £50 in the Ordinary account of a Bank giving $2\frac{1}{2}\%$ p.a. interest, and the rest in the Special Investment account giving $5\frac{1}{2}\%$ p.a.
   - *b* To place it all in a Building Society giving $4\frac{3}{4}\%$ p.a. interest.

6. A box of 48 packets of potato crisps is bought for 60p and the packets are sold at $2\frac{1}{2}$p each. Find the profit as a percentage of the selling price.

7. A dealer buys 150 kg of tea at 60p per kg and a further 50 kg at $52\frac{1}{2}$p per kg. He mixes them and sells the blend in $\frac{1}{4}$-kg packets at 18p per packet. Find his profit as a percentage of the selling price, to two significant figures.

8. A second-hand car dealer buys a car for £144. What selling price will give him:
   - *a* a profit of 150%
   - *b* a loss of 10%
   - *c* a profit of $8\frac{1}{3}\%$
   - *d* a loss of 100%?

9. When a man earned £17·50 per week he was able to save £2·50 per week. His wages were increased by 6% but his weekly expenditure increased by 12%. How much was he able to save now?

Revision Exercises on Chapter 2

10 In April 1966 the number of unemployed in Germany was 125 thousand and in Britain 330 thousand. In 1967 the numbers were, in Germany 625 thousand and in Britain 610 thousand. Calculate the increase in each country as a percentage of the 1966 figure, to the nearest whole number.

In Britain in 1967 there were 27 million wage-earners. Calculate the percentage of the wage-earners who were unemployed, to the nearest 0·1%.

11 The consumption of porridge in Scotland in 1964 was 7000 tonnes. By 1968 it had dropped by 23%. What was the consumption in 1968? If consumption continued to drop at the same rate (23% every four years) what was the consumption in 1972?

Revision Exercises on Chapter 2
Ratio and Proportion

Revision Exercise 2A

1 Writing paper costs 25p for 100 sheets. What should you pay for 180 sheets?

2 Writing paper costs 25p for 100 sheets. How many sheets should you get for 20p?

3 It takes 5 days for a squad of 24 pickers to gather a crop of raspberries. How many pickers, working at the same rate, would have gathered the crop in 3 days?

4 100 shares in an electric company cost £48.
  a How much would 175 shares cost?
  b If I want to invest £120 in the company how many shares will I get?

5 To insure a house for £2800 a man pays an annual premium of £5·60. What premium would his next-door neighbour require to pay at the same rate if his house is valued at £3300?

6 A vertical post $1\frac{1}{2}$ metres high casts a shadow 72 cm long. Find:
  a the length of the shadow cast by a tree 10 m high
  b the height of a flagpole which casts a shadow 10 m long (nearest $\frac{1}{10}$ m).

Arithmetic

7. In a schools' 11-a-side hockey tournament 22 goals were scored. How many goals would have been scored if each team was made up of 7 players?

8. A chicken farmer estimates that he has enough grain, etc., to feed his 2000 hens for a fortnight. If he acquires 800 more hens, for how long can his hens be fed?

9. To supply single portions of ice-cream for a school party of 54 pupils costs £2·16. To be on the safe side ice-cream worth £2·60 was ordered. How many pupils should be lucky enough to get a second helping?

10a. What is the R.F. of a map with a scale of 5 cm to 1 km?
  b. If two towns are 12·5 cm apart on this map, what is the actual distance between them?
  c. If the real length of a road is 9 km, what will its length be on the map?

11. £5 will buy 90 Danish kroner. How many kroner will you get for £8? How much British money will you get for 72 Danish kroner?

12. The graph in Figure 1 enables examination marks out of 140 to be

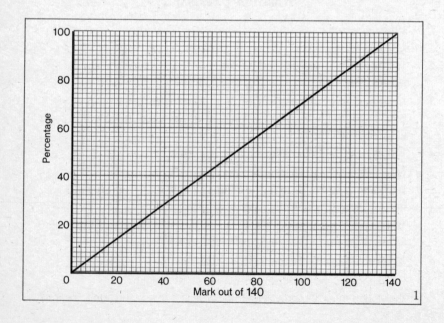

1

Revision Exercises on Chapter 2

read off as percentages. The original marks (out of 140) are shown on the horizontal axis and the percentages on the vertical axis. Read off as accurately as you can

a the percentages equivalent to 140, 100, 62 and 35 marks
b the marks out of 140 equivalent to 20%, 50%, 76%, 91%

Revision Exercise 2B

1 I paid the Town Council £65 last year on my rental of £52. What was the rate of payment per £ of rental? What did my neighbour pay whose rental was £72?

2 A cyclist, averaging 20 km per hour, takes 1 hour 12 minutes for a journey. How long would a car take to do the same journey at an average speed of 54 km per hour? (Answer to the nearest minute.)

3 If a certain sum of money is invested at 6%, it yields an income of £111. What income could be expected if the rate of interest fell to $5\frac{1}{2}\%$?

4 In a library it is estimated that 42 metres of shelving are required for every 1000 books. What length is required for 2250 books?

5 To supply 4 people, potatoes are boiled for 20 minutes. How long should they be boiled for 6 people?

6 A railway journey between two towns took $2\frac{3}{4}$ hours. If a new type of engine is used which increases the average speed of the train from 72 to $82\frac{1}{2}$ km per hour, find how many minutes are saved on the journey.

7 Out of each £1 of revenue from rates a city spends 36 pence on education and 28 pence on housing. If the total amount spent on housing is £504 000, how much is spent on education?

8 When I stand with both feet on a weighing machine my weight is shown as 60 kg.
   a What will the machine register if I stand with one foot on it and the other on the floor?
   b What will the machine register if I stand with one foot on it and hold the other foot in the air?

9 The wavelength of a radio transmission is inversely proportional to

| Arithmetic |

the frequency. Radio 2 has a wavelength of 1500 metres and a frequency of 200 kilocycles per second. Find:

a the frequency of the Hilversum transmission which has a wavelength of 400 metres

b the wavelength of Radio 4 which has a frequency of 809 kilocycles per second (answer to the nearest metre).

10 Copy and complete this exchange table:

British currency (£)	1	5	10		30
Belgian francs			600	3240	3600

Draw a graph for converting British and Belgian money. From it, find:

a how many francs can be obtained for 50p, £13, £21·50

b how many pounds can be obtained for 30, 1200, 3450 francs.

11 Draw a graph to convert examination marks out of 80 to percentages. Take 1 cm to represent 10 marks horizontally and 1 cm to represent 10% vertically. From your graph read off as accurately as you can:

a the percentages equivalent to 60, 36 and 55 marks

b the marks out of 80 corresponding to 50%, 72% and 43%.

Check your answers to *a* and *b* arithmetically.

12 On a certain Saturday in 1971 in the six principal soccer leagues there were 36 home wins (H), 28 away wins (A) and 12 draws (D). Find the ratios: *a* A:H   *b* H:D   *c* D:(A+H+D).

What percentage of all the games ended in draws? (Answer to nearest 1%.)

Revision Exercises on Chapter 3
Probability

Revision Exercise 3A

1 If a letter is chosen at random from the letters of the word FANTASTIC, what is the probability that it is:

a F   *b* A   *c* a vowel   *d* not a vowel?

Revision Exercises on Chapter 3

2 A shelf contains 65 copies of Book 1, 40 of Book 2, 35 of Book 3 and 60 of Book 4. If a book is chosen at random, what is the probability that it is:

   a Book 1   *b* Book 2   *c* Book 3 or Book 4?

3 The dial on a television set has five positions. Two give BBC1, two give BBC2 and one gives ITV. If a position is selected at random, what is the probability that it is:

   a ITV   *b* a BBC station   *c* BBC2?

4 *a* What is the probability that a card chosen at random from a pack will be a king?

   b How many kings would I expect to get if I picked a card 500 times, replacing the card after each selection?

5 Out of 300 football matches one year 56 were draws and 137 were home wins. Calculate, to 2 decimal places, the probability of a draw, of a home win and of an away win.

6 What is the probability of scoring a 4 when a die is rolled? How many times would I expect to have to roll the die to get a 4 turning up ten times?

   Repeat these questions for a die in the shape of a four-sided pyramid (tetrahedron) with the numbers 1, 2, 3, 4 printed on its sides.

7 Over a large number of football matches it was found that the probability of a draw was 0·23 and of an away win 0·34. How many draws and away wins might be expected on a Saturday with 55 matches?

8 Using the figures in question 7, how many home wins would be expected during a season when 1000 games were played?

9 Two towns, Aiton and Beeton, are joined by three roads X, Y and Z. On a journey from Aiton to Beeton and back the roads are chosen at random. What is the probability of X being used both ways?

10 The probability of certain seeds germinating is $\frac{3}{4}$, and the probability of a seed that has germinated producing a flower is $\frac{2}{3}$. If I buy a packet of 500 seeds, how many would I expect to germinate, and how many to produce flowers? What is the probability that a seed picked at random from the packet will produce a flower?

Arithmetic

Revision Exercise 3B

1. A bag contains 18 red beads, 24 blue beads and 36 white beads. If one is chosen at random, what is the probability that it is:
   - *a* red
   - *b* blue
   - *c* red or blue?

2. A grocer estimates that 5% of the eggs he receives are cracked. What is the probability that one of the eggs selected at random will be cracked? How many cracked eggs would he expect in a batch of 50 dozen?

3. When I throw darts at a target I find that the probability of hitting the target is 0·8. When I do hit the target the probability that I will score a bull's-eye is 0·1. Out of 100 throws, how many bull's-eyes might I expect?

4. 5000 tickets were sold in a raffle. If I buy 5 tickets, what is the probability that I will win first prize? How many tickets would I have to buy to raise this probability to $\frac{1}{10}$?

5. If I roll two dice 108 times, how often should I expect to get a score of:
   - *a* 12
   - *b* 1
   - *c* 7?

6. If I toss three coins simultaneously 100 times, how often may I expect to get:
   - *a* three Tails
   - *b* 2 Tails and one Head, in any order?

7. A regular tetrahedron has the numbers 1, 2, 3, 4 on its faces. If it is rolled twice and the number on the base noted each time, what is the probability of getting a total:
   - *a* greater than 6
   - *b* less than 3
   - *c* greater than 6 or less than 3?

   What is the most likely total for two rolls, and what is the probability of this total?

8. Eight men and eight women enter for a mixed-doubles tennis tournament. What is the probability that Mr A will be drawn to partner Miss B?

9. Two brothers enter a singles knock-out tennis tournament. If there are 16 entries, what is the probability that the brothers will meet in the first round?

10. A bag contains 3 white marbles and 2 red ones. Two marbles are drawn from the bag simultaneously. What is the probability that:

a they are both red   *b* one is white and one is red?

If a marble is drawn out and replaced, and another is then drawn out, what is the probability that both are white?

Revision Exercise on Chapter 4
Time, Distance, Speed

Revision Exercise 4

1. During the summer holiday season a special train leaves London at 11 30 on Sunday and arrives in Rome at 13 30 on Monday. How long does the journey take? If the train arrives in Genoa at 07 50 on the Monday how long does the journey from London to Genoa take?

2. A train leaves Glasgow at 23 07 and arrives in Largs, 49 km away, at 00 17. Find the time for the journey, and the average speed in km/h.

3. A train leaves Dunkirk at 22 35 and arrives in Paris, 264 km away, at 02 20. Find its average speed.

4. The Forth Road Bridge is 2·4 km long; how long will it take to cross:
  a in a car at an average speed of 48 km/h,
  b walking at an average speed of 4·5 km/h?

5. When a boy walks at an average speed of 5 km/h he takes $2\frac{1}{4}$ hours for a journey. How long would he take if he cycled at 15 km/h? What is the length of the journey?

6. On a flight from London to Cairo a plane leaves London at 22 30 and arrives in Cairo at 07 00 next morning. The distance from London to Cairo via Rome, where the plane stopped for 1 hour, is 3870 km. Find the average speed of the plane while it is in the air.

7. A train leaves Basle at noon and is timetabled to reach Stuttgart, 259 km away, at 15 30. What is its average speed?

   On a certain day it travels at this speed for 2 hours and is then delayed for half an hour. If it can only average 30 km/h for the rest of the journey, when does it arrive in Stuttgart?

Arithmetic

8 A space-craft is launched towards the planet Mars and travels at an average speed of 16000 km/h. If the distance between Mars and the Earth is about 72 million km, show that it will take about six months to get there.

9 A motorist takes 2 h 45 min for a certain journey when he travels at an average speed of 84 km/h. How long will he take for the journey if his average speed is 90 km/h? Do this question:

  a by proportion   *b* by finding the length of the journey.

10 A slow train leaves Stirling for Perth, 56 km away, at noon. Its average speed when moving is 60 km/h. It stops for 4 minutes at Dunblane (10 km from Stirling) and for 8 minutes at Gleneagles (32 km from Stirling). By drawing a graph on 5-mm squared paper, find when the train reaches Perth. Take 5 squares horizontally to represent 10 minutes and 5 squares vertically to represent 10 km.

On the same diagram draw the graph for an express train which leaves Perth at 1210 for Stirling, moving at 90 km/h with one stop of 2 minutes at Gleneagles. Hence find when the express reaches Stirling, and when and where the two trains pass one another.

Answers

Answers

Answers

Algebra—Answers to Chapter 1

Page 4 Exercise 1

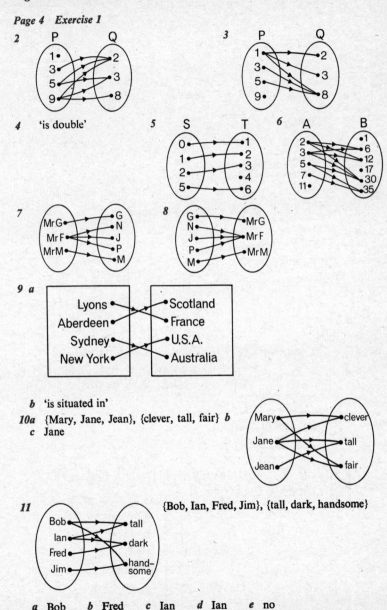

4 'is double'

9 b 'is situated in'

10 a {Mary, Jane, Jean}, {clever, tall, fair} *b*
c Jane

11 {Bob, Ian, Fred, Jim}, {tall, dark, handsome}

a Bob *b* Fred *c* Ian *d* Ian *e* no

Answers

Page 6 Exercise 2

1 *a* {(J, F), (T, F), (T, C), (B, F), (B, T), (B, C), (I, T)}

b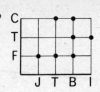

2 *a*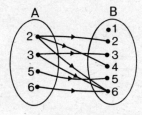

b {(2, 2), (2, 4), (2, 6), (3, 3), (3, 6), (5, 5), (6, 6)}

3 {−1, 1, 3, 5, 7}, {2, 4, 6, 8, 10}; 'is 3 less than'

4 *a* {(0, 0), (2, 1), (4, 2), (6, 3), (8, 4)}   *b*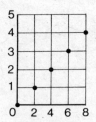

5 {(*a, n*), (*a, s*), (*a, t*), (*i, f*), (*i, n*), (*i, s*), (*i, t*), (*o, f*), (*o, n*), (*u, s*)}

6 *a* 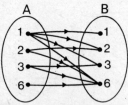   *b* {(1, 1), (1, 2), (1, 3), (1, 6), (2, 2), (2, 6), (3, 3), (3, 6), (6, 6)}

7 *a* {(1, 2), (1, 3), (1, 4), (1, 5), (2, 3), (2, 4), (2, 5), (3, 4), (3, 5), (4, 5)}
b {(1, 1), (2, 2), (3, 3), (4, 4), (5, 5)}
c {(2, 1), (3, 1), (3, 2), (4, 1), (4, 2), (4, 3), (5, 1), (5, 2), (5, 3), (5, 4)}

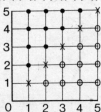

Answers

8 a Rows: $(-1, 1), (-1, 2), (-1, 3), (-1, 4); (0, 1), (0, 2), (0, 3), (0, 4); (1, 1), (1, 2), (1, 3), (1, 4)$ **b**

9 $\{(p, q), (q, p), (p, r), (r, p), (p, s), (s, p), (q, r), (r, q), (q, s), (s, q), (r, s), (s, r)\}$

10a $\{\frac{1}{2}, 1, 2, 4, 8\}, \{8, 4, 2, 1, \frac{1}{2}\}$. The first number multiplied by the second number gives 4.

b

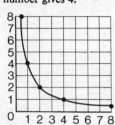

Page 10   Exercise 3

1 (*i*), (*iv*) and (*vi*)

2 a   **b** $\{(a, 1), (b, 2), (c, 2)\}$

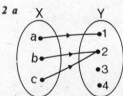

3 a $\{(a, 3), (b, 2), (b, 4), (c, 1)\}$
b *b* is linked to *two* elements of *Y*.

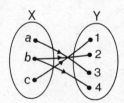

4 a $\{(1, 8), (3, 8), (2, 9), (4, 9)\}$
b yes

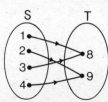

Answers

5 a {(3, 1), (6, 2), (9, 3), (12, 4), (15, 5), (18, 6)}
  b No; not all elements of S are included.
6 a 1, 1, 1, 2, 1, 3   b f = {(1, 1), (2, 1), (3, 1), (4, 2), (5, 1), (6, 3)}
7 {(1, 3), (3, 5), (5, 7), (7, 9), (9, 11)}
8 a 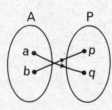   b Reverse the direction of the arrows.

9 A → X, B → Y, C → Z; A → X, B → Z, C → Y; A → Y, B → X, C → Z;
   A → Y, B → Z, C → X; A → Z, B → X, C → Y; A → Z, B → Y, C → X

10a There are 8 mappings   b 9 mappings

Page 12 Exercise 4

1 (*i*) and (*iii*)   2a   b

3 a the numbers of copies, and the costs   b £6·30; 10
4 a the distances flown, and the times taken   b 2000 km; ½ hour
5   *a*: *c*: *d*
7 a (*1*) M/OP (*2*) I/NA   b (*1*) £2·60 (*2*) £1·14
8   THEY SEE ME
10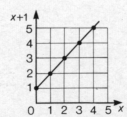

11b (*i*) THIS IS GOOD FUN (*ii*) ALL SYSTEMS GO

Page 15 Exercise 5

1 a
x	0	1	2	3	4
$x+1$	1	2	3	4	5

b and c

Answers

2 a

x	0	1	2	3	4
2x	0	2	4	6	8

x	0	1	2	3	4
3x	0	3	6	9	12

x	0	1	2	3	4
4x	0	4	8	12	16

b

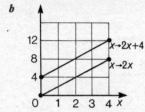

c straight lines; through the origin; the greater the coefficient of x the greater the slope of the line

3 a

x	0	1	2	3	4
2x	0	2	4	6	8

x	0	1	2	3	4
2x+4	4	6	8	10	12

b

c straight lines; parallel; only one through origin

4 a {(0, 0), (1, 1), (2, 4), (3, 9), (4, 16)}   **b**

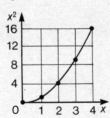

5 a, b

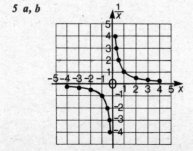

Answers

6 c The graphs of $x \to x$ and $x \to -x$ pass through the origin and are perpendicular to each other if the scales are the same on both axes. The graphs of $x \to x+2$ and $x \to -x+2$ pass through the point (0, 2) and are perpendicular to each other.

7 b For A, $x \to x+1$; for B, $x \to 1-x$

Algebra—Answers to Chapter 2

Page 18 Exercise 1A

1	1	2	10	3	−3	4	0
5	−1	6	−4	7	0	8	−6
9	−6	10	−2	11	0	12	−6
13	−8	14	−2	15	2	16	2
17	−2	18	−7	19	0	20	5
21	−8	22	0	23	0	24	−18
25	3a	26	0	27	−3c	28	−4c
29	−2d	30	−12p	31	−7q	32	−3r
33a	a	b	−a	34	identity element		

Page 19 Exercise 1B

1 −2   2 −5   3 each is 8   4 each is −11
5 6   6 −2   7 −3   8 8   9 2
10 11   11 −21   12 −5   13 −21   14 0
15 −2y   16 4z   17 0   18 0   19 −6c
20 8   21 −5   22 −2x   23 −13y   24 4a+b
25 −8p

Page 20 Exercise 2

1 5a   2 6b   3 4c   4 4d   5 −2e
6 −4f   7 −5g   8 −5h   9 14x   10 0
11 0   12 −14x   13 4x+2y   14 −p−6q   15 2k
16 −4m−n   17 2a+4c   18 4a−6b   19 −p−q−r   20 0

Page 21 Exercise 3

1 a 4   b 14   c −14   d −4   2 8   3 2
4 −4   5 10   6 −6   7 16   8 0   9 −4
10 −1   11 1   12 −8   13 3   14 −3   15 1
16 −4   17 −11   18 6   19 8   20 −8   21 3
22 36   23 −12   24 −12   25 0   26 −8   27 8

Answers   233

| 28 | 8 | 29 | −8 | 30 | T | 31 | T | 32 | F | 33 | T |
| 34 | T | 35 | F | 36 | F | 37 | T | | | | |

Page 22   Exercise 4

1	$7x$	2	$2y$	3	$3y$	4	$-z$	5	$-6a$	6	$-2b$
7	$2c$	8	0	9	0	10	$-12p$	11	0	12	$13r$
13	$-x$	14	0	15	$-5z$	16	$-2a$	17	$-6b$	18	$-9c$
19	$3a$	20	$-2b$	21	$9c$	22	$8d$	23	$-7e$	24	$-13f$
25	g	26	$-3h$	27	0	28	$14x$	29	$-14x$	30	0
31	$2x+6y$			32	$5p$	33	$-2l$	34	$7n$	35	$2b-2c$
36	$8c$	37	$3p+3q$	38	$4x+6y+8z$						

Page 23   Exercise 5

1   a square   *a*   (1, 0), (−2, 0), (−2, −3), (1, −3)
                b   (0, 1), (−3, 1), (−3, −2), (0, −2)
2   0; (3, −8), (−3, −8)   3   900 m   4   230   5   11°C; −9°C
6   110 m; 15 m; 125 m

Page 23   Exercise 5B

1   400 m   2   £230   3   3, −2, −7, −12, −17, −22
4   18, 12, 2, −12, −30, −52
5 *a*   The bird is flying 1 m/s south relative to the ground.
   b   +10 m/s; −11 m/s; −1
6 *a*   (1)   4°C   (2)   13°C   (3)   6°C
   b   (1)   8°C   (2)   7°C   *c*   noon to 3 pm; 6 to 9 pm

Page 25   Exercise 6

1   0, −4, −8, −12, −16, −20   2   0, −5, −10, −15, −20, −25

3–5	25	20	15	10	5	0	−5	−10	−15	−20	−25
	20	16	12	8	4	0	−4	−8	−12	−16	−20
	15	12	9	6	3	0	−3	−6	−9	−12	−15
	10	8	6	4	2	0	−2	−4	−6	−8	−10
	5	4	3	2	1	0	−1	−2	−3	−4	−5
	0	0	0	0	0	0	0	0	0	0	0
	−5	−4	−3	−2	−1	0	1	2	3	4	5
	−10	−8	−6	−4	−2	0	2	4	6	8	10
	−15	−12	−9	−6	−3	0	3	6	9	12	15
	−20	−16	−12	−8	−4	0	4	8	12	16	20
	−25	−20	−15	−10	−5	0	5	10	15	20	25

6	−15	7	−15	8	4	9	−4	10	4	11	6
12	0	13	0	14	1	15	0	16	−16	17	9
18	−2	19	25	20	25	21	−20				

Answers

Page 26 Exercise 7

1	0	**2**	0	**3**	40	**4**	40	**5**	−40	**6**	−40
7	100	**8**	−18	**9**	−14	**10**	−14	**11**	14	**12**	14
13	−54	**14**	12	**15**	−12	**16**	−100	**17**	100	**18**	56
19	−36	**20**	64	**21**	$2a$	**22**	$-3b$	**23**	$4c$	**24**	$-5d$
25	x^2	**26**	x^2	**27**	$-5y$	**28**	$2y$	**29**	$-2p$	**30**	$3q$
31	$-4r$	**32**	s^2	**33**	4	**34**	−4	**35**	1	**36**	0
37	5	**38**	−1	**39**	0	**40**	−10	**41**	−2	**42**	0
43	1	**44**	−4	**45**	4	**46**	−4	**47**	0	**48**	1
49	−9	**50**	7	**51**	−1	**52**	−10				

Page 27 Exercise 8

1 −2 **2** 0 **3** 2 **4** 3 **5** −4 **6** −6
7 0 **8** 1 **9** 4 **10** 4 **11** 9 **12** −3
13 −5 **14** −4 **15** 0 **16** −19 **17** −2 **18** 0
19 12 **20** 4 **21** −1 **22** 0 **23** −8 **24** 81
25 a, c, f **26** {4, −4} **27** {1, −1} **28** {2, −2} **29** ø **30** {2}
31 {−3} **32** {−1} **33** {4} **34** {1, −1} **35** {0} **36** {10, −10}
37 {−5} **38** T **39** F **40** T **41** F **42** F
43 T **44a** 20 m/s **b** 35 m/s **45a** (*1*) 10 m/s (2) −20 m/s
b The stone is moving (*1*) up (2) down.

Page 28 Exercise 8B

1 −50 **2** 0 **3** 60 **4** 126 **5** −1 **6** 30
7 {9, −9} **8** ø **9** 0 **10** {−6} **11** {10} **12** {3, −7}
13 {−3, −11} **14** {3, −1} **15** {−2, −1, 0, 1, 2}
16 {−1, −2, −3, ...} **17** Z **18** {5, 6, 7, ...} **19** T **20** T
21 F **22** T **23a** 10 m **b** 10 m **c** 0 m
d −20 m. The stone is below the point of projection **24a** 26°C; 14°C
b 14°C, 26°C. The temperature is falling 3°C per hour.
25 900, 600, −100; profits of £900 and £600 and a loss of £100
26 −1, 0, 1, 2, 3, 4, 5 **27** 2

Page 30 Exercise 9B

1 a 60 **b** −60 **c** −120 **d** 28 **2** 30 **3** 42
4 40 **5** 0 **6** −20 **7** −24 **8** 20 **9** 36
10 −25 **11** −40 **12** −54 **13** 36 **14** −64 **15** −60
16 18 **17** 24 **18** −180 **19** −36 **20** 18 **21** 1
22 4 **23** −4 **24** −2 **25** T **26** F **27** T
28 T **29** T **30** T

Answers 235

Page 31 Exercise 10

1 Rows 5 −2   3   1   5;   −10 15   5
          −2   5 −4   1   −2;   −10   8 −2
          −6 −3 −2 −5   30;   18 12 30

2 Rows −3   5 −1   6 −18;   −15   3 −18
            4   2 −5   7   28;   8 −20   28
           −1 −5 −2 −3   3;   5   2   3

Page 32 Exercise 11

1 $7x-14$ **2** $-2a-12$ **3** $-3x+3$ **4** $5x-5$
5 $-6x+6y$ **6** $8a+8b$ **7** $-16+4b$ **8** $3x+15$
9 $ax-2bx$ **10** $px-qx$ **11** $-2az+bz$ **12** $-x^2-xy$
13 $-15x+12$ **14** $-10m+15n$ **15** $2ap+2aq$ **16** $7x-5x^2$
17 $-x+y$ **18** $-2x+3y$ **19** $-5+4x$ **20** $-3a-2b$
21 $4a-12b+8c$ **22** $-3a-9b+15c$ **23** $-5+5x+5x^2$
24 $-4x^2+6x-8$ **25** $-x^2+2x+3$ **26** $-5+4x+3x^2$

Page 32 Exercise 12

1 $7a$ **2** $-7a$ **3** $-3y$ **4** $4z$ **5** $-6x$ **6** $-5q$
7 $-4p$ **8** 0 **9** a **10** $-9b$ **11** $15c$ **12** $-18k$
13 0 **14** $-5n$ **15** $-x+2y$ **16** $-5x+y$ **17** $8x+2y$
18 $-a+5b$ **19** $-2a-b$ **20** $2a-3b$ **21** $2x+5y$ **22** $-a+8b$
23 $-3a+3b$ **24** $3p-2q$ **25** $7x-2y$ **26** $3x+2y$ **27** $3x-3y$
28 $4+3x$ **29** $7+5x$ **30** $-2x+8$ **31** $3-2x$ **32** $3x+3$
33 $4+x$ **34** $6x+1$ **35** $2x-3$ **36** $3+8x$ **37** $5x+y$
38 $x+7y$ **39** $4x-4y$ **40** -1

Page 33 Exercise 12B

1 $4x^2-7y$ **2** $-7a^2-3b^2$ **3** x^2-x **4** $x+5y$
5 $x+5y$ **6** $4x-6y$ **7** $6a-2b$ **8** $-3x-y$
9 $-5a-b$ **10** $2x^2-5x+3$ **11** $4y^2-3y-7$ **12** $5-7z+2z^2$
13 $-x-7y$ **14** $7x$ **15** $2x^2+2x-3$ **16** $14y^2-19y$
17 0 **18** 0 **19a** $3x+3y+3z$ **b** $-x+y-3z$
c $-8x+7y+z$

Page 35 Exercise 13

1 $\frac{6}{8}, \frac{75}{100}$ **2** $\frac{-2}{4}, \frac{3}{-6}$ **3** $\frac{8}{10}, \frac{16}{20}$ **4** $0, \frac{0}{2}$ **5** $\frac{3}{1}, \frac{15}{5}$ **6** $\frac{5}{4}$
7 $\frac{1}{4}$ **8** $-\frac{1}{4}$ **9** $-\frac{5}{4}$ **10** 0 **11** $\frac{22}{15}$ **12** $\frac{2}{15}$
13 $-\frac{2}{15}$ **14** $-\frac{22}{15}$ **15** 0 **16** $\frac{1}{6}$ **17** $-\frac{1}{12}$ **18** $-\frac{9}{8}$
19 $-\frac{3}{4}$ **20** $-\frac{3}{2}$ **21** $\frac{1}{5}$ **22** 1 **23** $\frac{5}{4}$ **24** $\frac{1}{8}$
25 1 **26** $-\frac{5}{8}$ **27** $-\frac{3}{2}$ **28** $-\frac{1}{2}$ **29** $\frac{3}{4}$ **30** $-\frac{3}{4}$

Answers

Page 35   Exercise 14

1	$\frac{6}{5}$	2	$\frac{3}{4}$	3	$\frac{1}{2}$	4	$\frac{2}{1}$ or 2	5	$\frac{3}{2}$	6	$-\frac{3}{2}$
7	$\frac{3}{5}$	8	$-\frac{7}{4}$	9	$-\frac{3}{1}$ or -3	10	$\frac{1}{4}$	11	$\frac{1}{5}$	12	$-\frac{1}{2}$
13	$\frac{4}{1}$ or 4	14	$\frac{8}{3}$	15	none	16	$-\frac{5}{8}$	17	-1	18	none
19	2	20	4	21	6	22	10	23	1	24	$\frac{1}{5}$

Page 37   Exercise 15

1	-5	2	-3	3	4	4	-2	5	-6	6	-1
7	9	8	-8	9	-7	10	-3	11	-8	12	4
13	$\frac{4}{3}$	14	$-\frac{4}{3}$	15	-8	16	12	17	$\frac{3}{5}$	18	-3
19	$-\frac{3}{2}$	20	$\frac{1}{2}$	21	$\frac{7}{6}$	22	0	23	0		
24	not possible			25	-2	26	4	27	-5	28	$\frac{1}{3}$
29	6	30	-6	31	-6	32	6	33	-2	34	-3
35	7	36	-4	37	-1	38	0	39	0	40	$-\frac{1}{3}$

Page 37   Exercise 16B

1	$4+2a$	2	$3c+4$	3	$4x-6y$	4	$5+2m$	5	$x+2$
6	$3p+9q$	7	$8a-4b$	8	$c+3$	9	$x-y$	10	$1+a$
11	$4+5y$	12	$2c-5$	13	$-a-3b$	14	$-2p+3$	15	$3t-4$
16	$-2c-3$	17	$-4x-5$	18	$-2y+1$	19	$2-x$	20	$5-a$
21	$-p-2q$	22	$-2+2z$	23	$-3x-3y$	24	$-p-4q$	25	$-a-2b$
26	$-3+x$	27	$1-y$	28	$a-b$	29	$-7+3y$	30	$1+3y$
31	$\frac{1}{2}$	32	-1	33	1	34	$\frac{3}{2}$	35	$\frac{2}{3}$
36	0	37	-2	38	impossible				

Algebra—Answers to Chapter 3

Page 40   Exercise 1

1 all true   2 both true   3b $x = 4$, {4}   c yes
4 b $x = 10$, {10}   c yes

Page 42   Exercise 2

(In this, and similar exercises, the variables are omitted to save space.)

1	11	2	12	3	-3	4	-7	5	7	6	54
7	10	8	20	9	-4	10	4	11	-10	12	8
13	4	14	3	15	1	16	-5	17	7	18	-3
19	50	20	5	21	12	22	0	23	-20	24	-13

Page 42   Exercise 2B

| 1 | {1} | 2 | {$\frac{1}{4}$} | 3 | {$\frac{3}{4}$} | 4 | {$-\frac{1}{4}$} | 5 | {$\frac{1}{4}$} | 6 | {$\frac{1}{4}$} |
| 7 | {$\frac{3}{2}$} | 8 | {$\frac{1}{4}$} | 9 | {-1} | 10 | {3} | 11 | {$\frac{1}{5}$} | 12 | {0} |

236

Answers

Page 42   Exercise 3
1   all true   2   both true   3b $x = -6, \{-6\}$ c   yes
4 b $x = \frac{15}{2}, \{\frac{15}{2}\}$ c   yes

Page 43   Exercise 4
1 $\frac{1}{2}$ 2 $\frac{1}{5}$ 3 $\frac{8}{5}$ 4 2   5 $-\frac{1}{2}$ 6 $-\frac{1}{13}$ 7 $-\frac{8}{5}$
8 -2 9 $\frac{1}{4}$ 10 $-\frac{1}{4}$ 11 1   12 -1 13 $\frac{4}{3}$ 14 $-\frac{4}{3}$ 15 $-\frac{5}{6}$
16 $\{5\}$ 17 $\{6\}$ 18 $\{1\}$ 19 $\{\frac{2}{3}\}$ 20 $\{-8\}$ 21 $\{-\frac{3}{7}\}$ 22 $\{\frac{1}{2}\}$
23 $\{0\}$ 24 $\{-\frac{1}{3}\}$ 25 $\{\frac{3}{5}\}$ 26 $\{-2\}$ 27 $\{-\frac{1}{3}\}$ 28 $\{\frac{1}{4}\}$ 29 $\{\frac{2}{9}\}$
30 $\{\frac{2}{3}\}$ 31 $\{\frac{1}{12}\}$ 32 $\{3\}$ 33 $\{12\}$ 34 $\{4\}$ 35 $\{-10\}$ 36 $\{20\}$
37 $\{21\}$ 38 $\{-2\}$ 39 $\{0\}$

Page 44   Exercise 5
1 3   2 $\frac{1}{2}$ 3 7   4 1   5 7   6 $\frac{7}{5}$ 7 -4
8 $\frac{8}{3}$ 9 $-\frac{2}{3}$ 10 0   11 $\frac{5}{2}$ 12 -1 13 $\frac{1}{8}$
14 $\frac{1}{3}$ 15 9   16 6   17 5   18 6   19 $-\frac{3}{2}$
20 -10 21 $-\frac{1}{2}$ 22 3   23 1   24 $\frac{2}{3}$

Page 45   Exercise 5B
1 $\{18\}$ 2 $\{3\}$ 3 $\{4\}$ 4 $\{\frac{11}{2}\}$ 5 $\{-2\}$ 6 $\{0\}$
7 $\{5\}$ 8 $\{-3\}$ 9 $\{\frac{9}{4}\}$ 10 $\{\frac{4}{7}\}$ 11 $\{-\frac{1}{3}\}$ 12 $\{\frac{15}{4}\}$
13 $\{-\frac{4}{3}\}$ 14 $\{\frac{12}{7}\}$

Page 46   Exercise 6
1 a–c   all true   2   both true   3a $\{-4, -3, -2, -1, 0, 1, 2\}$
  b $x < 3; \{-4, -3, ..., 2\}$ c   yes
4 a $\{5, 6, 7, 8, 9, 10\}$ b $x \geqslant 5; \{5, 6, ..., 10\}$ c   yes

Page 47   Exercise 7
1   $x > 3$ 2 $y < 4$ 3 $z > -3$ 4 $p \leqslant 5$
5   $t \geqslant -5$ 6 $y < 14$ 7 $m < 7$ 8 $x > 14$
9   $v < 5$ 10 $x < -6$ 11 $y > -10$ 12 $x \geqslant 12$
13   $y < 12$ 14 $x \geqslant 15$ 15 $m > -1$ 16 $x > 0$
17   $x > 4$ 18 $y < 0$

Page 48   Exercise 8
1 a $\{4, 5, 6\}$ b $\{2, 4\}$ c $\{7, 9, 11\}$
  d $\{4\}$ e $\{2, 4, 6, 8\}$ f $\{3, 5, 7, 11, 13\}$
2   $\{y: y < 26\}$ 3 $\{p: p < -3\}$ 4 $\{x: x > -42\}$
5   $\{z: z \geqslant -9\}$ 6 $\{m: m > 39\}$ 7 $\{t: t \leqslant 8\}$
8   $\{t: t \geqslant 0\}$ 9 $\{y: y < 2\}$ 10 $\{p: p > \frac{7}{5}\}$
11 $\{x: x > 9\}$ 12 $\{x: x > 7\}$ 13 $\{y: y < -10\}$
14 $\{m: m < 1\}$ 15 $\{n: n \leqslant 9\}$ 16 $\{x: x > -6\}$

Answers

Page 48   Exercise 9

1 *a* all true except (*4*), (*5*) and (*7*)   *b* all true except (*4*), (*5*) and (*7*)
2 *b* (*1*) yes   (*2*) no   (*3*) yes   3*b* (*1*) yes   (*2*) no   (*3*) yes
4 *b* (*1*) yes   (*2*) no   (*3*) yes

Page 50   Exercise 10

1. $x > 2$
2. $y < 6$
3. $z \leqslant 1$
4. $x > 0$
5. $t < 5$
6. $x > 4$
7. $x \geqslant 1$
8. $t < \frac{3}{4}$
9. $w < \frac{3}{2}$
10. $x < 12$
11. $x > -3$
12. $y \leqslant -\frac{2}{3}$
13. $x > -3$
14. $z < -5$
15. $x > 0$
16. $x > -5$
17. $x > 5$
18. $z > -\frac{6}{5}$
19. $x \geqslant 4$
20. $y < -\frac{9}{4}$
21. $\{x: x > 10\}$
22. $\{y: y < 4\}$
23. $\{z: z > 3\}$
24. $\{x: x \leqslant 8\}$
25. $\{p: p > 9\}$
26. $\{x: x \leqslant 8\}$
27. $\{x: x > \frac{3}{4}\}$
28. $\{y: y < \frac{2}{3}\}$
29. $\{y: y \leqslant 3\}$
30. $\{p: p \leqslant 20\}$
31. $\{x: x > -7\}$
32. $\{m: m < -12\}$
33. $\{n: n < -1\}$
34. $\{t: t < -3\}$
35. $\{x: x < -7\}$
36. $\{y: y > -3\}$
37. $\{x: x < 3\}$
38. $\{x: x < 0\}$
39. $\{x: x \leqslant -100\}$
40. $\{y: y \leqslant 0\}$

Page 51   Exercise 11

1. $\{x: x > 7\}$
2. $\{y: y < -5\}$
3. $\{x: x > 2\}$
4. $\{y: y < 4\}$
5. $\{t: t \geqslant 4\}$
6. $\{x: x < 2\}$
7. $\{m: m > 3\}$
8. $\{x: x < 3\}$
9. $\{x: x \leqslant -2\}$
10. $\{t: t < -1\}$
11. $\{z: z > -8\}$
12. $\{t: t > 1\}$
13. $\{x: x < 0\}$
14. $\{y: y \geqslant 3\}$
15. $\{x: x \leqslant 20\}$
16. $\{x: x > 3\}$
17. $\{y: y \leqslant -5\}$
18. $\{z: z > \frac{5}{2}\}$

Page 51   Exercise 11B

1. $\{m: m < -1\}$
2. $\{x: x > -\frac{3}{2}\}$
3. $\{p: p > \frac{7}{2}\}$
4. $\{y: y > \frac{1}{2}\}$
5. $\{x: x \leqslant -4\}$
6. $\{y: y < 4\}$
7. $\{z: z < -1\frac{3}{2}\}$
8. $\{p: p < -8\}$
9. $\{x: x \leqslant 1\}$
10. $\{x: x > -\frac{1}{2}\}$
11. $\{x; x < \frac{4}{3}\}$
12. $\{x: x < 1\}$
13. $\{x: x < 5\}$
14. $\{x: x > \frac{9}{2}\}$
15. $\{x: x > 2\cdot 8\}$

Page 53   Exercise 12

1. $\{12\}$
2. $\{30\}$
3. $\{\frac{5}{2}\}$
4. $\{x: x < -\frac{3}{4}\}$
5. $\{x: x > -9\}$
6. $\{x: x < -2\}$
7. $\{m: m > -4\}$
8. $\{20\}$
9. $\{t: t > 6\}$
10. $\{-28\}$
11. $\{s: s < 12\}$
12. $\{7\}$
13. $\{x: x \leqslant 9\}$
14. $\{11\}$
15. $\{y: y > \frac{1}{8}\}$
16. $\{y: y > \frac{5}{4}\}$
17. $\{t: t > \frac{4}{7}\}$
18. $\{-2\}$

Answers

Page 53 Exercise 12B

1 {1}   *2* {x: x > 3}   *3* {n: n < 11}   *4* {0}
5 {t: t ⩾ 6}   *6* {y: y > 10}   *7* {½}   *8* {m: m < 3/5}
9 a {12}   *b* {6}   *c* {10/3}   *d* {−6}   *e* {15/2}
10a {−9}   *b* {x: x > 4}   *11a* {6/5}   *b* {4/3}   *12* {−½}
13 {x: x ⩽ 17/26}

Page 54 Exercise 13

1 12   *2* 5   *3a* (4x+6) mm   *b* 6   *c* 9 mm, 6 mm
4 a (6x−8) cm   *b* 9   *c* 13 cm, 10 cm, 130 cm²
5 a (3y−1) cm   *b* 8   *c* 8 cm, 10 cm, 5 cm
6 a (2x+90)°   *b* x = 45; 90°, 60°, 30°
7 a 9x°   *b* x = 20; 80°, 61°, 39°   *8* x > 7/2   *9* x > 7
10 1; 12 cm   *11* x > 10 (also x > 1½ from length (2x−3) m)
12 3·5   *13* 3   *14* 6 mm, 12 mm, 12 mm
15a P = {x: x > −2}, K = {x: x < 5}   *b* {−1, 0, 1, 2, 3, 4}
16 28, 29   *17* −38, −39   *18a* 24, 25, 26   *b* no
19a 34, 36, 38   *b* 19, 21, 23   *20* 4p, 8p

Page 56 Exercise 13B

1 a (8a+8) m   *b* 4x m   *c* x = 2a+2   *d* the square   *2* 9p, 27p
3 7x = 70; 10p, 20p   *4* {x: x > 3/2, x ∈ Q}, {x: x < 6, x ∈ Q}
5 60+40x = −20; −2 points   *6* x = 6; 144 m²   *7a* 19−x, 17−x
b 9   *8a* 27−x, 14−x   *b* 11
9 (i) x < 7, x > 1   (ii) y > 4, y < 12   (iii) z > 5, z < 17
10a ½(2x−10)+2(2x+10) km   *b* 25
11a 2x km, 3/2(x−10) km   *b* (*1*) 46 km/h   (*2*) 54 km
12 ½ hour, or 30 min

Algebra—Answers to Revision Exercises

Page 60 Revision Exercise 1

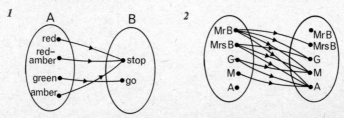

Answers

3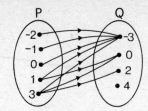

4 a {(−1, −1), (0, 0), (1, 1), (2, 2), (4, 4)}
  b {(−1, 1), (0, 0), (1, 1), (2, 4)}
  c {(1, −1), (2, 0), (4, 2)}

5 a $A = \{2, 5, 8, 11, 14, 17, 20\}$ *b* $B = \{1, 4, 7, 10, 13, 16, 19\}$ and
  $C = \{3, 6, 9, 12, 15, 18\}$ *c* The required set is S

6 a {(2, 0), (2, 2), (2, 4), (2, 6), (3, 0), (3, 3), (3, 6), (4, 0), (4, 4), (6, 0), (6, 6)}

b 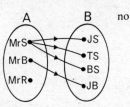 *7a* (ii), (iii), (v), (vi) *b* (v)

8 a 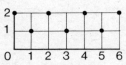 no *b* 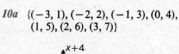 yes

9 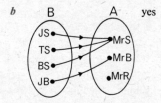 *10a* {(−3, 1), (−2, 2), (−1, 3), (0, 4), (1, 5), (2, 6), (3, 7)}

b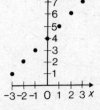

c $a = 21$, $b = -14$, $c = 2\frac{1}{4}$

Answers

11a (i), (ii) and (iii) **b** (ii)

12 $\frac{1}{2}$ $\frac{4}{5}$ $\frac{3}{20}$ $5\frac{1}{4}$
    ↕   ↕   ↕   ↕
    0·50 0·80 0·15 5·25

13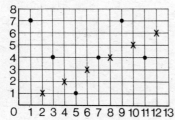

14 b and c

15a {(−4, −4), (−2, −3), (0, −2), (2, −1), (4, 0), (6, 1) (8, 2), (10, 3)}

b, c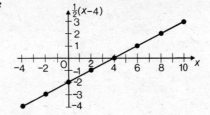

16a {(0, 12), ($\frac{1}{2}$, 8), (1, 6), (2, 4), (3, 3), (5, 2), (7, 1$\frac{1}{2}$), (11, 1)}

b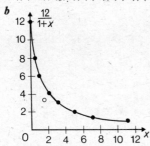

17a Monday and Fridays **b** Wednesdays and Saturdays **c** Fridays

Page 63 Revision Exercise 2A

1 a 13 **b** 3 **c** −3 **d** −13 **e** 11c
 f 3c **g** −3c **h** −11c **2a** 289 **b** 7°C
 c −£0·50 **3a** 2p **b** $3x − 6y$ **c** 3b **d** $−3u + v$

4 a 5 **b** −10 **c** −14 **d** 6 **e** 0
 f −2p **g** −7x **h** 5m **5a** $7c + 2d$ **b** 0
 c $8x − 2y$ **6a** $x + 4y$ **b** $4a − b$ **c** $−2p − 2r$ **d** $−2y + 5z$

Answers

7   60; −60; 36; 28   8a   35   b   −56   c   −36
   d   20   e   −2p   f a^2 g   2a   h $-5x^2$
9 a   15   b   −7   c   0   d   −5   e   4
   f   0   g   1   h   0
10a $-2a-6b$ b $10a-15b$ c $-12x+15$ d $2p+3q$ e $2x-4$
    f $5a-14$
11a $-\frac{1}{8}$ b $-\frac{5}{6}$ c $\frac{1}{8}$ d $-\frac{4}{15}$ e   5
    f   −4   g   −4   h $\frac{6}{5}$

Page 64   Revision Exercise 2B

1 a   8a   b   0   c   −6n   d   −4m   2a   T
    b   F   c   T   d   F   3a   −11, −16, −21
    b   −24, 48, −96   c   −13, −19, −26
    d   120, −720, 5040   4a $2x; -2y+2z$
    b $2b-5c; 4a+4b-3c$ c $3p+10q+8r; -p+10r$
    d $-2x+4; 2x^2+6x-2$ 5a   126   b   −120   c   −27
    d   0·25   6a   −5   b   24   c   −2   d   5
    e   3   f   16   7a $-10u+6v$ b   3c   c   3x
    d   −4a   e $-3x^2+4$ f $-2p+7q$ g $-4a+2b$ h $-3a+8b$
8 a   15, 7, −1, −9, −17; −137   b   yes. 17   9   x can be 0, 1, 2, 3, 4, 5;
    y can be 2, 3, 4; 55 m²   10   0, −5, −8, −9, −8, −5, 0; {−2, 4}
11a $-\frac{11}{35}$ b $\frac{1}{12}$ c $-\frac{2}{3}$ d $-\frac{7}{10}$
12   $p = 15-a; a > 5$ 13a   16   b $\frac{7}{2}$ c $3x+2$.   13
14a   £$\frac{1}{20}p$ b   £$\frac{19}{20}p$ c   £$\frac{19}{2}p$

Page 66   Revision Exercise 3A

1 a   {14}   b   {−6}   c   {−1}   d   {7}   e   {−8}   f   {8}
2 b $9 > 5 \Leftrightarrow$ B' is to the right of A'   c $1 > -3 \Leftrightarrow$ B'' is to the right of A''
    d   They are of equal length
3 b   (1) $6 < 10$ (2) $-3 > -5$ (3) $-6 > -10$
4 a   {$\frac{4}{3}$}   b   {$m: m > 12$}   c   {$y: y \geq \frac{1}{6}$}   d   {$\frac{3}{4}$}
    e   {$x: x > -6$}   f   {$x: x \leq 12$}   g   {$\frac{2}{5}$}   h   {$p: p < 0$}
5 a   {$x: x > 3$}   b   {$x: x > 6$}   c   {$x: x \leq -5$}
    d   {$x: x < 2$}   e   {$x: x < 3$}   f   {$x: x \leq 1$}
6 a   T   b   F   c   T   d   T
7 a   {19}   b   {$x: x < -2$}   c   {$y: y < \frac{1}{3}$}
    d   {$\frac{27}{2}$}   e   {$t: t \leq \frac{6}{5}$}   f   {$t: t < -9$}
8   5   9a $k+2, k+4$ b   24, 26, 28   10   67

Page 67   Revision Exercise 3B

1 a   {13}   b   {−2}   c   ∅   d   {0}   e   {4}   f   {3}
2 a   {$x: x < 2$}   b   {$x: x < -8$}   c   Q   d   {$y: y < -\frac{1}{36}$}
    e   {$m: m > \frac{5}{24}$}   f   {$t: t \leq 3$}
3 a   T   b   F   c   F   d   T   4 $10y-5$ a $1\frac{3}{4}$ b $\frac{1}{2}$

Answers

5 a $\{a: a < -8\}$ b $\{2\}$ c $\{y: y < 2\}$ d $\{0\cdot10\}$
  e $\{a: a \geqslant 6\}$ f $\{y: y \geqslant -\frac{1}{30}\}$
6 a $\{x: 2 < x < 10\}$ b $\{x: -9 < x \leqslant 3\}$ c $\{x: 2 < x < 6\}$
  d $\{x: -2 < x \leqslant 1\}$
7 $\{-2, -1, 0, 1, 2, 3, 4\}$; $\{x: -3 < x < 5\}$
8 $1\frac{1}{2}$ hours; 12 km 9 $\frac{1}{8}x + \frac{1}{10}(37-x) = 4$; model in question *8*
10 $4(x+10) + 8x = 12x+40$; $12x+40 \leqslant 100 \Leftrightarrow x \leqslant 5$
11 11 12a 10·5 litres b $\dfrac{1050}{14+x}$ c $0\cdot7(14+x) = 10\cdot5, \Leftrightarrow x = 1$

Geometry—Answers to Chapter 1

Page 73 Exercise 1

2 b 180°, 360°, etc. c the coloured dot

Page 74 Exercise 2

1 b (iv) and (v)

3
a	b	c	d	e	f	g	h	i	j	k	l
2	0	1	1	4	1	4	1	2	0	1	1
yes	no	no	no	yes	no	yes	no	yes	no	no	no

4 a {A, E, W} b {H} c {H, N, Z}
5 a Rows: 4, 2; 8, 4; 2, 1; 6, 3
  b The numbers in the second last column are double those in the last column.

Page 77 Exercise 3

5 b Draw a line from each vertex or corner, perpendicular to the dotted line, and produce it its own length to the image vertex; join these images.

Page 80 Exercise 4

1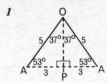

2a 10 cm, 2·7 cm, 4 cm b 90°, 55°, 125°
3a A b B′ c AB′ d B′B
  e ∠AB′B f ∠B′AY

4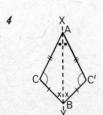

5a 7 cm, 10 cm, 70° b XY, BB′
6 A′B′ is equal and parallel to AB; A′B′ does not meet XY.

Answers

Page 81   Exercise 4B

1

2a As P is on XY and B′ is the image of B in XY, PB = PB′.
b AP+PB = AP+PB′ = AB′. Because APB′ is a straight line.

3 Join M to the image of N in XY, cutting XY at the required point.

4 (2·4, 10), (0, 4), (1·6, 0), (6, 5·5)

5 PQ is the image of BD in AC. If O is mid-point of AC,
AO = QO = BO = CO = PO = DO = ½AC. O.

6 AB is parallel to XY.

Page 83   Exercise 5

1 b (1, −6), (2, −2), (5, −4)   **c** (−1, 6), (−2, 2), (−5, 4)

2 (3, −3), (−1, −2), (7, −2); (−3, 3), (1, 2), (−7, 2)

3 a (0, 5) ↔ (0, −5), (1, 6) ↔ (1, −6), (3, 8) ↔ (3, −8), (5, 10), ↔ (5, −10), (8, 0) ↔ (8, 0)   **b** (8, 0)

4 a (0, 5) ↔ (0, 5), (1, 6) ↔ (−1, 6),   (3, 8) ↔ (−3, 8),   (5, 10) ↔ (−5, 10) (8, 0) ↔ (−8, 0)   **b** (0, 5)

5 (4, −1), (4, −3), (0, −3), (0, −1); (−4, 1), (−4, 3), (0, 3), (0, 1)

6 a (11, 4), (10, 2), (9, 4), (8, 5)   **b** (−5, 4), (−6, 2), (−7, 4), (−8, 5)

7 Rows: (0, 0), (2, −2), (3, 0), (0, 2), (−4, 4); (0, 0), (−2, 2), (−3, 0), (0, −2), (4, −4); (6, 0), (4, 2), (3, 0), (6, −2), (10, −4)

Page 84   Exercise 5B

1 a (9, 2)   **b** (11, −1)   **c** (−2, 0)   **d** (1, −4)

2 (12, 1), (9, 1), (9, 3), (12, 3)

3 a (13, 2)   **b** (−3, −2)

4 a (1, −2), (3, 1), (6, 0), (−3, 4)   **b** (1, 0), (3, 3), (6, 2), (−3, 6)

5 P(a, b) → P′(a, 2k−b)

6 a (0, 5), (6, 6), (1, 4), (8, 1), (−2, 2)   **b** (0, −5), (−6, −6), (−1, −4), (−8, −1), (2, −2)

Page 86   Exercise 6

2 (6, 5), (3, 5)   **3** (0, −2), (0, 1)   **4**

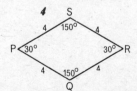

Answers

245

5 *a* 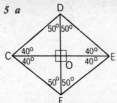   *b* CD = DE = EF = FC
CO = OE, DO = OF

6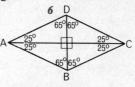

7 *b* The sum of the angles of the rhombus = the sum of the angles of the two triangles = $180° + 180° = 360°$

8 *a* T   *b* F   *c* T   *d* T   *e* F   **9** 48 square units

10 $p = 5$ **11** *a* 12 cm^2 *b* 80 cm^2

12 A ↔ C, B ↔ D, AB ↔ CD, BC ↔ DA. AB is parallel to DC, BC to AD.

Page 91   Exercise 10

1 B ↔ D, A ↔ A, C ↔ C, ∠ABC ↔ ∠ADC, △CDA ↔ △CBA

2   **3**

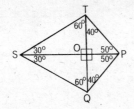

5 *a* F   *b* T   *c* T   *d* F   *e* T   **6** Rows: 4, 2; 4, 2; 2, 1

7 (0, 0), (2, 3), (4, 0), (2, −5)   **8** (0, 0), (−2, −3), (−4, 0), (−2, 5)

9 *a* a kite   *b* 64, 20 square units   *c* 24 square units

10 *a* 30 cm^2 *b* 135 cm^2

11 *a* T   *b* T   *c* F   *d* T   *e* T

Page 93   Exercise 11B

2 Two angles of 36° and two angles of 144°

3 AA′ is equal and parallel to BB′; AB is equal and parallel to A′B′; all the angles are right angles. ABB′A′ is a rectangle.

4 (0, 6), (2, 6)

Geometry—Answers to Chapter 2

Page 96   Exercise 1

1 *a* A ↔ C, B ↔ D, AB ↔ CD, AD ↔ CB
   b DC, BC   *c* DC, BC

Answers

2 a Q ↔ S, R ↔ P, QP ↔ SR, QR ↔ SP **b** PQ, RC

3 2 **4** H, I, O, S, X, Z **5a** H, J, F, O **b** EG, OH, IJ, GJ

Page 97 Exercise 2

1 b $(-3, -1), (-2, -5), (-6, 0), (0, 2), (-4, -4), (0, 0)$

2 $(-2, -1), (3, 3), (-1, 4), (6, 0), (-a, -b)$

3 $(1, 3), (3, 6), (1, 5), (7, 7), (8, 0)$

4 **5**

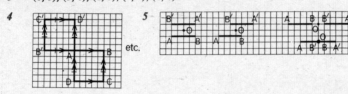

etc.

6 a, b **7**

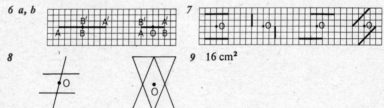

8   **9** 16 cm²

Page 99 Exercise 2B

1 a $(-5, -5), (-10, 1), (0, 12), (-p, -q)$ **b** 0

2 $P = \{H, I, N, X\}, Q = \{A, D, E, H, I, M, T, X\}, P \cap Q = \{H, I, X\}$

3 **4**

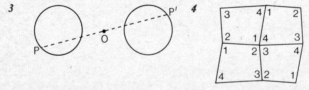

Page 100 Exercise 3

1 a A ↔ C, B ↔ D, △ABC ↔ △CDA **b** AB ↔ CD, AB = DC, AB ∥ DC
  c BC ↔ DA, BC = AD, BC ∥ AD
  d ∠ABC ↔ ∠CDA, ∠ABC = ∠CDA
  e ∠ACB ↔ ∠CAD, ∠BAC ↔ ∠DCA, ∠BCD = ∠DAB

2 a 180° **b** 360° **c** 180° **3** 2

4 a **b** a rectangle

Answers

Page 101   Exercise 4

1 (i)   △ABC ↔ △CDA
  (ii) AB = DC, AD = BC, AB ∥ DC, AD ∥ BC, ∠BAD = ∠DCB,
     ∠ABC = ∠CDA
  (iii) OA = OC, OB = OD

2 RS = 8 cm, SP = 5 cm, ∠R = 60°, ∠Q = ∠S = 120°

3 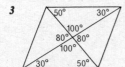   **4a** 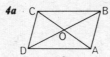   **b**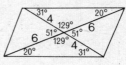

5 EF ∥ HG, EH ∥ FG, EF = HG, EH = FG, ME = MG, MF = MH
6 (1, 5), (4, 3)   **7** (3, 4), (−1, 4)
8 T, F, T, T, F, F, T, T   **9** T, F, F, T, T, F, T, T
10 T, T, F, T, T, F, T, T   **11a** a right angle   **b** adjacent sides equal

Page 102   Exercise 4B

1 half turn symmetry
2 midpoint of LP; parallelogram; half turn symmetry
3 a (9, 3)   **b** (−5, 3)   **c** (−1, −9)   **4** half turn symmetry
5 a half turn symmetry   **b** half turn symmetry

Page 104   Exercise 5

1 A → B, D → C, C → F, AD → BC, BC → EF, DC → CF,
    parm ABCD → parm BEFC
2 A → D, D → H, C → G, AD → DH, BC → CG, DC → HG,
    parm ABCD → parm DCGH
4 (6, 4)
5 a parallel sides   **c** (10, 4), (15, 6), (20, 8), (2) and (3)
6 a (3, 5)   **b** (8, 11)   **c** (80, 78)   **d** (1, 4)
7
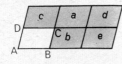

Page 106   Exercise 6

1 a 5   **b** 6   **c** 7   **d** 8
2 a ∠ADE = ∠ABC, ∠AED = ∠ACB   **b**

Answers 248

3 a ∠QDC = ∠QAB, ∠CBP = ∠DAP   *b*

4

5 a angles FMN, FGH; FNM, FHG   *b* △FMN   *c*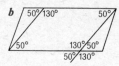

6 a ∠PTS, ∠TQU; ∠QUR, ∠TSU; ∠VUS, UQT   *b*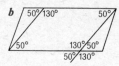

Page 108   Exercise 7

1 a (*1*) 6   (*2*) 3   *b* (*1*) 1, 5; 2, 6; 3, 7; 4, 8   (*2*) 1, 3; 2, 4; 5, 7; 6, 8
(*3*) 1, 2; 2, 3; 3, 4; 4, 1; 4, 5; 3, 6; 2, 5; 1, 6; etc.
(*4*) 1, 3, 5, 7   (*5*) 2, 4, 6, 8

2 a ∠FBC, ∠BCD; ∠EBC, ∠BCH; ∠BED, ∠EDK; ∠GED, ∠EDC
b

3 b ∠OAB, ∠OCD; ∠OAD, ∠OCB; ∠OBA, ∠ODC; ∠OBC, ∠ODA
c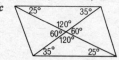

4

5 a ∠EAB, ∠ECD; ∠EBA, ∠EDC   *b*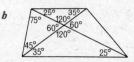

Answers

Page 109   Exercise 7B

1 *a* ∠DAE, ∠ABC; ∠EAC, ∠ACB

2   ∠ABC = 70° (corr. ∠s), ∠ACE = 45° (alt. ∠s),
  ∠ACB = 65° (∠BCD = 180°)

3

```
   /80°
  /100°    130°\
 /80°      50°  \
/100°      130°  \
/80°        50°   \
```

5 *a* (*1*) t, w (2) u, z (3) $s, q+r$ (4) $v, p+q$ *b*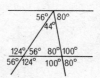

Page 110   Exercise 8

1 (i) 24 cm²   (ii) 35 cm²   (iii) 144 cm²   (iv) 16 cm²
2 opposite sides parallel, 18 sq units
3 *b* 312 mm²   **4** parm, 19 sq units   **5** 12 sq units

Geometry—Answers to Chapter 3

Page 115   Exercise 1

1 *d* on the circumference of the circle   *e* an infinite number
  f {P: CP ≤ 2 cm}   *g* {P: CP ≥ 2 cm}   *h* {P: CP = 2 cm}

2 *a*   *b*   *c*   *d*

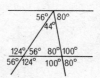

3 *a*   *b*   *c*   *d*

4   **5**

6 *a* Surface of sphere, centre O, radius 6 cm
  b interior of sphere   *c* space outside sphere

Answers

250

Page 116 Exercise 2

1 *b* (*1*) the *y*-axis
      (*2*), (*3*), (*4*) lines parallel to *y*-axis, 1, 2 and 3 units to the right of it
      (*5*) the region to the right of the *y*-axis

2 *b* (*1*), (*2*), (*3*) lines parallel to the *y*-axis, 1, 2 and 3 units to the left of it
      (*4*) the region to the left of the *y*-axis

Page 117 Exercise 3

1 *a*   *b*   *c*   *d*

2 *a* the *y*-axis   *b* the *x*-axis

3   **4**   **5**

6 *a*   *b*   *c*   *d* (continued)

7 *a*   *b*   *c*   *d*

8   (i) {P: OP < 20}   (ii) {(*x*, *y*): 0 ⩽ *x* < 10}
        (iii) {P: 8 ⩽ OP ⩽ 15}   (iv) {(*x*, *y*): −8 < *y* ⩽ 12}

Answers

Page 118 Exercise 3B

1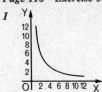

2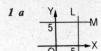

3   the perpendicular bisector of AB

Page 120 Exercise 4A

1 a   *b* $L \cap M = \{(5, 5)\}$

2 a   *b* $L \cap M = \{(-2, -1)\}$

3 a   *b* $L \cap M = \{(0, 2)\}$

4 a
b $(2, 6), (6, 6), (2, -1), (6, -1)$
c $x = 6, x = 2; y = 6, y = -1$

5 a

b $P \cap Q = \{(3, -2)\}, Q \cap R = \{(-4, -2)\}, R \cap S = \{(-4, 5)\},$
$S \cap P = \{(3, 5)\}$ *c* $P \cap R = \emptyset, Q \cap S = \emptyset$

6 a   *b* 49 square units

7 a   *b* 16 square units

Answers

Page 121 Exercise 4B

1 a   **b** $L \cap M = \{(3, -1)\}$

2 a   **b** $L \cap M = \{(-5, 0)\}$

3 a   **b** $(4, 4), (4, -2), (-2, -2)$

4 a   **b** $A \cap B = \emptyset$, $A \cap C = \{(6, 3)\}$
$A \cap D = \{(6, -1)\}, B \cap C = \{(2, 3)\}$
$B \cap D = \{(2, -1)\}, C \cap D = \emptyset$

5 a   **b** 8 square units

6 a   **b** 20 square units

7 a ...

8 a

b $P \cap Q = Q, Q \cap R = Q, P \cap R = Q$

Page 123 Exercise 5B

1 circle parallel to the equator

Answers

253

2 (great) circle passing over north and south poles   3 a square

4 a ——————   b ⊚   c ▭   d ▭

5 a   arc of a circle, centre O, radius 100 cm

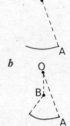

b

Page 124   Exercise 6

1 b $y = x$ 2 A, C, D, E, O
3 6, −7, 1000, −15, 0, 0·1   4c (2), (4), (5)
5 P, S, T, O   6 −10, 2, −1, −1000, 0, 25
7 c above: D, E, F; below: A, B, G   d $p < q$
8 b $p < -q, q < -p$ 10

Page 127   Exercise 7

1   (0, 0), (0, 1), (0, −1)   2   3

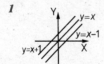

4 Parallel lines through a (0, 0), b (0, 100), c (0, −10); all at 45° to x-axis

5   6   7   8

Answers

Page 128   Exercise 8

1   **2**   **3**   **4**

5 $1, 2, 3, \frac{1}{3}$

6 a $y = x$ b $y = 5x$ c $y = -x$
   d $y = \frac{1}{10}x$ e $y = -10x$ f $y = 123x$

7   **8**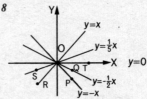

Page 129   Exercise 9

1   **2**   **3**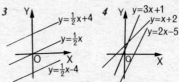

4 (included in above)

5 a $-1, (0, 0)$ b $-1, (0, 3)$ c $-2, (0, -4)$ d $-3, (0, 1)$

6

Page 130   Exercise 10

1   **2**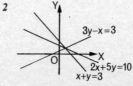

Answers

3 *a*   *b*   *c*   *d*

4 *a* $A \cap B = B$ *b*   *c* $A \cap D = \emptyset$ *d*

Geometry—Answers to Revision Exercises

Page 133 Revision Exercise 1A

2 *a* 6 cm, 10 cm   *b* 60°, 90°   3 60°
4   triangles APP′, AQQ′, OPP′, OQQ′   5 6 cm, 60°
6 *a* (0, 0), (5, −5), (10, 0), (5, 5)
  b (0, 0), (−5, 5), (−10, 0), (−5, −5)
7   (5, 6), (8, 2), (7, −3), (4, 0), (1, 1), (−4, 3); (15, 6), (12, 2), (13, −3), (16, 0), (19, 1), (24, 3). *x*-coordinates increased by 12, *y*-coordinates equal.
8   Each side is 8 cm long; ∠QPS = 70°, ∠PQR = ∠PSR = 110°
10   121°
12   a kite; 32 square units   13 (6, −4); 40 square units
14*a* (4, −3), (8, −5), (12, −3), (8, −1)
  b (−4, 3), (−8, 5), (−12, 3), (−8, 1)
15   The altitude and bisectors are concurrent.

Page 135 Revision Exercise 1B

1   bilateral: (i) 1, (ii) 8, (iii) 6, (iv) 5. Half turn: (ii) and (iii)
2 *a* isosceles △ *b* isosceles △ *c* kite   *d* rhombus
3   (5, 8), (14, 4), (20, 0), (*a*+10, *b*)
4   (5, −8), (−4, −4), (−10, 0), (−*a*, −*b*)
5 *a* line through O at 45° to *x*-axis *b* (0, 0), (0, 4), (2, 4), (8, 0), (8, 6), (*q*, *p*)
6   angles NPM′ and PM′M   7 10 cm
8   (10, 6) (15, 8), (20, 10), (25, 12), *c*
10 a square; 8   11*a* no   *b* yes   *c* $c \in S$

Page 136 Revision Exercise 2A

1   2 midpoint of AC or BD; on CD 6 cm from D; on AB 2 cm from A

Answers

3 $(-3, -5), (2, -4), (-1, 5), (-a, -b), (p, -q)$

4 $(1, -2), (4, 0), (7, 1), (8, 5)$ **5** H, I, N, O, S, X, Z

6 *a* same straight line   *b* parallel straight line   **7a** 7 cm   *b* 30 cm²

8 *a* T   *b* T   *c* F   *d* T   *e* F   *f* F   *g* T

9 midpoint of diagonal, 2

10a T   *b* T   **11**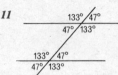

12 $x = z, y = u, x+y = 180, x+u = 180, z+u = 180, y+z = 180$

13a ∠APQ, ∠PQD; ∠BPQ, ∠PQC
   b ∠APQ, ∠BPQ; ∠CQP, ∠DQP; ∠APQ, ∠CQP; ∠BPQ, ∠DQP

14 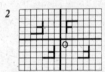   **15**   **16** 20 sq units

Page 138   Revision Exercise 2B

1 Z, N, X, H, O, S

2

3 *a* $(-12, -8)$ *b* $(-3, -6)$ *c* $(-11, -2)$,   *d* $(-a-6, -b-4)$

4 $(-3, 1), (2-p, -2-q)$

5 *a* midpoint of AB   *b* midpoint of AC; slide without turning

6 16 cm²

7 *a* rectangle, square, rhombus, parallelogram
   b rectangle, square, rhombus, parallelogram
   c square, rhombus   *d* square   *e* rectangle, square
   f square, rhombus   *g* kite   *h* square, rhombus

8 *b*

c 3, 3, 2

Answers

10 (i) 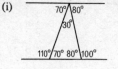 (ii) 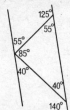 **11**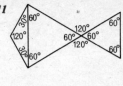

12a **b** **c** $x+y = 90$. $\angle \text{LOM} = 90°$

13 slide without turning; subtract equal triangles from whole figure; breadth of rectangle = height of parallelogram; base × height

14 $(q-p)(r-1)$ sq units

Page 140 Revision Exercise 3A

1 a **b** **c** **d**

2 **3a** **b**

4 (i) $\{(x,y): 2 < y < 5\}$ (ii) $\{(x,y): -2 \leqslant x \leqslant 5\}$ (iii) $\{P: OP \leqslant 5\}$

5   points which are 5 cm from two vertices

6   the empty set

7 $\{(5, 3)\}$

8 (i) points 4 cm from O and 3 cm from AB
(ii) points 3 cm from P and 4 cm from CD

Answers

9   **10**   **11**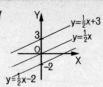

12 both *a* and *b*

Page 142 Revision Exercise 3B

1 *a*   *b*   *c*   *d*

2 $\{(x, y): 3 \leqslant x \leqslant 5 \text{ and } -1 \leqslant y \leqslant 2\}$
points inside and on the boundary of the rectangle with sides $x = 3$, $x = 5$, $y = 2$, $y = -1$

3 $\{(x, y): 0 < x \leqslant 3 \text{ and } y = 4\}$
points with *x*-coordinates greater than zero but less than or equal to 3, and with *y*-coordinate 4

4   (i) $\{(x, y): -4 \leqslant x \leqslant 3 \text{ and } 2 \leqslant y \leqslant 6\}$
     (ii) $\{P(x, y): \text{OP} < 4 \text{ and } y \geqslant 0\}$
     (iii) $\{(x, y): x > 0, y > x \text{ and } y < 3\}$

5   points less than 5 cm from all three vertices

6   **7**

8   (i) points less than 7 cm from A and also less than 9 cm from B
     (ii) locus of centre of a wheel of radius 2 cm which rolls round the outside of a rectangle 10 cm long and 5 cm broad

Answers

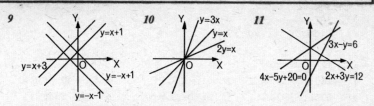

12a the surface of a sphere, centre O, radius 5 cm
  b the surface of a cylinder with radius 5 cm, and axis the given line
  c two parallel planes 5 cm from the given plane

Arithmetic—Answers to Chapter 1

Page 148 Exercise 1

1 £2·20   2 £12·60   3 £16·18   4 £28·21
5 £38·84   6 £22·05

Page 149 Exercise 2

1 £13·25   2 £23·75   3 £6·53   4 £50·90
5 £17·28, £31·68, £300·96

Page 150 Exercise 3

1 60p, 65p, £2·40, £3·50
2 a 77p   b £1·08   c £4·50   d £3·04
3 a 49p, £4·90, £49   b 72p, £7·20, £72
4 7 two-star (1p); 7 three-star (no change); 6 four-star (5p); 6 five-star (2p)
5 14 two-star (2p); 13 three-star (6p); 13 four-star (2p); 12 five-star (4p)
6 a £3·75   b £11·25   c £22·50   d £37·50   e £132   f £92·25
7 a £2·85   b £3·42   c £4·27½   d £8·55   e £9·50   f £47·50
  g £475   h £12·25½

Page 151 Exercise 4

1 a £8   b £2   c £2·92   d 22p
2 £75, £13·50, £76, £24, 51p, 76p   3 £10·11
4 a £48·75   b £19·50   c £30·81
5 a £1·65   b £3·75   c £10·87½   d £75
6 £186·73   7a £32   b £56   c £133·34   d £97
8 £251·57   9 £129·35   10 £402·19

Page 154 Exercise 6

1 £15   2 £8   3 £2·31   4 £7

Answers

5 42p   **6** £25   **7** £15   **8** £5
9 £4·50   **10** £7500   **11** £1 312 500

Page 155 Exercise 7
1 £6; £156·20   **2** £11; £231·75   **3** £26·40; £686·40
4 £0·72; £13·22   **5** £6·60; £126·60   **6** £1·37; £56·75
7 £3; £153   **8** £25; £1275   **9** £1·62; £28·62
10 £23·50; £2373·50   **11** £8·47; £366·07   **12** £1·64; £76·64

Page 156 Exercise 8
1 8%   **2** $6\frac{1}{4}$%   **3** $7\frac{1}{2}$%   **4** $12\frac{1}{2}$%   **5** 4%   **6** 8%
7 26%   **8** 9%   **9** $6\frac{1}{4}$%   **10** $7\frac{3}{4}$%

Page 157 Exercise 9
1 £7 profit; £5 loss; 18p profit; no profit or loss; £163 profit
2 £200   **3** £1·50   **4** £1·70 profit
5 £5·04 profit   **6** £1·95 profit   **7** £4·45

Page 159 Exercise 10
1 20% profit   **2** 10% loss   **3** $12\frac{1}{2}$% profit   **4** 100% profit
5 20% profit   **6** $12\frac{1}{2}$% loss   **7** $12\frac{1}{2}$% loss   **8** $33\frac{1}{3}$% profit
9 *a* 20% profit   *b* $16\frac{2}{3}$% profit   **10a** 25%   *b* $33\frac{1}{3}$%
11 $16\frac{2}{3}$% profit   **12** $16\frac{2}{3}$% profit

Page 160 Exercise 11
1 £16·50   **2** 78p   **3** 90p   **4** 96p   **5** 38p
6 6p   **7** £572   **8** £118·80   **9** $37\frac{1}{2}$p
10 £57·60   *a* $16\frac{2}{3}$%   *b* 14%   **11** 4·5p; £2·$62\frac{1}{2}$; 14·3%

Page 161 Exercise 12
1 $\frac{1}{4}$, 0·25, 25%   **2** $\frac{1}{2}$, 0·5, 50%   **3** $\frac{1}{5}$, 0·2, 20%
4 $\frac{1}{8}$, 0·125, 12·5%   **5** $\frac{4}{5}$, 0·8, 80%   **6** $\frac{7}{20}$, 0·35, 35%
7 $\frac{1}{10}$, 0·1, 10%   **8** $\frac{7}{10}$, 0·7, 70%   **9** $1\frac{1}{2}$, 1·5, 150%
10 $\frac{8}{1000} = \frac{1}{125}$, 0·008, 0·8%

Page 162 Exercise 13
1 *a* 33·3%   *b* 12%   *c* 16·7%   *d* 60%   *e* 8·5%
2 76%, 75%, 87·5%   **3** 20 cm². 20%, 15%, 25%, 40%
4 80%   **5** £13·97   **6a** $12\frac{1}{2}$%   *b* 20%   *c* $3\frac{1}{3}$%   *d* 5%
7 *a* 15p   *b* $2\frac{1}{2}$p   *c* $62\frac{1}{2}$p

Answers

8 27000, 19200, 13800, 7800   **9** 1·5 kg, 1·05 kg, 0·45 kg
10 70   **11** 180   **12** 160   **13** £24·50   **14** £3460
15 £9·68, £9·97

Arithmetic—Answers to Chapter 2

Page 165   Exercise 1

1 a 5:8   b 4:3   c 9:5   d 2:5   e 3:5   f 5:2
2 a 1:4   b 3:2   c 3:4   d 5:3   e 12:11   f 2:3
3 a 3:2   b 2:3   **4** 1:1   **5** the *actual* dimensions
6 a 10 cm   b 7·5 cm   **7** a 5:4   b 5:4   c 25:16
8 a 5:4   b 4:5   c 5:9   **9** 4:5
10a (*1*)135   (*2*)198; decreased, increased
   b (*1*)dec.   (*2*)inc.   (*3*)no change   (*4*)dec.   (*5*)inc.   (*6*)inc.   (*7*)dec.
11a (*1*)£40   (*2*)£28   (*3*)£33   b   45 cm   (*2*)50 cm   (*3*)35 cm
   c (*1*)6 m   (*2*)5¼ m   (*3*)4½ m   d   £1·68   (*2*)£1·80   (*3*)£2·31
   e 99
12a (*1*)160   (*2*)196   b (*1*)68   (*2*)135

Page 168   Exercise 2

1 a 3   b 4½   c 39   d 5   e 10   f 56
2 15   **3** 18   **4** 50p   **5** £4   **6** 29p, £3·19
7 3p, £1·80   **8** 285 km   **9** £30·24   **10** 750, 150
11 72000, 72

Page 170   Exercise 3A

1 £1·16   **2** 99p   **3** 80p   **4** 684 km
5 a 48   b 10 kg   **6** £33·20   **7** 87½p
8 £18·40   **9** 70%   **10**a 15·2 m   b 17·5 cm

Page 170   Exercise 3B

1 63p   **2** £1·26   **3** 1·95 m   **4** £4500
5 3 cm   **6** £15·36   **7** 91   **8** £9·90
9 200 m, 125 or 126 steps   **10** £92·40

Page 172   Exercise 4

1 a no   b 10p or 11p; 280p or a little less   **2** no   **3** £1·35
4 £4·20   **5** —   **6** 6·75 m   **7** £91   **8** —
9 135   **10** 1280   **11**a £155   b £960
12a 9·72m   b 8

Answers

Page 174   Exercise 5

1 *a* 1·54 cm   *b* 63·2 cm   *c* 0·26 cm
2 *a* 120 m   *b* 13·4 m   *c* 2·85 m
3 *a* 1 m   *b* 1 : 100   *c* 3·65 m   *d* 27·8 cm
4 *a* 1 : 200000   *b* 47 cm   *c* 54·8 km
5 *a* 84 cm   *b* 5·9 km   6 *a* 7·6 km   *b* 7·4 cm
7 *a* 34 m by 18·5 m, 629 m²   *b* 25 m², 800 m²
8 *a* 61·2 cm   *b* 81·6 cm

Page 177   Exercise 6A

1 *b, c, e, f*   2 20 km/h   3 40   4 60 km/h
5 32   6 80   7 5   8 $2\frac{1}{2}$ weeks

Page 178   Exercise 6B

1 70 km/h   2 36   3 9·6 h, 384000 km
4 90 km   5 6 hours   6 5
7 *a* 1·2, 4, 30 ohms   *b* 1·2, 3, 0·5 amps
8 *a* 90, 45, 112·5 cm³   *b* 1, $1\frac{7}{8}$, 0·45 atmospheres

Page 179   Exercise 7

1 *a* 19, 48, 75, 94 pence   *b* 16, 22, 43, 54 schillings
2   number of km: 15, 30, 45, 60, 75, 90, 105, 120
  a 6·7, 5·7, 3·3 litres   *b* 37·5, 82·5, 102 km
3   premiums (£): 1, 2, 3, 4, 5, 6. £3·75, £4·50, £5·25, £5·63
4   weights (kg): 1, 2, 3, 4, 5, 6
  a 1·6, 2·4, 4·4, 5·5 kg   *b* 75, 125, 225 m²
5   time (h): 40, 30, 24, 20, 15, 12. 38·4, 12·8
6 *a* 56, 83, 89 pence   *b* $3\frac{1}{2}$, 10, 13 kroner

Page 181   Exercise 8A

1 105 kg   2 65   3 38 min   4 700
5 £110   6 $7\frac{1}{2}$ min   7 £12·50   8 20 cm, 8 cm
9 16 cm, 10 cm; 160 cm²   10*a* 16 cm   *b* 12 km

Page 182   Exercise 8B

1 1500   2 200   3 17·6 min   4 1·25 m
5 £54   6 *a* 32 cm   *b* 8·6 km
7 *a* 3 : 2   *b* 9 : 4   *c* 27 : 8   8 8
9 0·945 km²   10 2 hours 5 minutes

Answers

Arithmetic—Answers to Chapter 3

Page 190   Exercise 2A

1 a $\frac{1}{5}$ *b* $\frac{2}{5}$ *c* $\frac{1}{5}$ *d* $\frac{3}{5}$ *2 a* $\frac{1}{3}$ *b* $\frac{2}{3}$; 1
3 a $\frac{5}{7}$ *b* $\frac{2}{7}$; 1   *4 a* $\frac{1}{2}$ *b* $\frac{1}{4}$ *c* $\frac{2}{5}$ *5 a* $\frac{1}{6}$ *b* $\frac{1}{2}$
6 a $\frac{1}{13}$ *b* $\frac{1}{52}$ *c* $\frac{1}{4}$ *d* $\frac{3}{13}$ *7 a* $\frac{1}{2}$ *b* $\frac{1}{3}$ *c* $\frac{1}{6}$ *d* $\frac{1}{2}$
8 a $\frac{3}{10}$ *b* $\frac{1}{5}$ *c* $\frac{1}{2}$ *d* $\frac{1}{2}$ *9 a* $\frac{1}{5}$ *b* $\frac{1}{4}$ *10 a* $\frac{1}{2}$ *b* $\frac{1}{3}$

Page 192   Exercise 2B

1 a $\frac{4}{11}$ *b* $\frac{4}{11}$ *c* $\frac{3}{11}$ *2 a* $\frac{1}{4}$ *b* $\frac{1}{2}$ *3* $\dfrac{1}{2\,000\,000}$
4 a $\frac{3}{5}$ *b* $\frac{1}{6}$ *5 a* $\frac{1}{9}$ *b* 0   *c* $\frac{1}{3}$ *d* $\frac{4}{9}$
6 a $\frac{13}{20}$ *b* $\frac{7}{20}$ *c* $\frac{2}{5}$ *d* $\frac{3}{5}$ *e* $\frac{11}{20}$ *f* $\frac{1}{4}$ *g* $\frac{7}{20}$ *h* $\frac{1}{5}$
7 $\frac{3}{5}, \frac{1}{2}$ *8* $\frac{1}{4}, \frac{8}{11}$

Page 194   Exercise 3

1 50   *2 a* 10   *b* 20   *c* 40   *3 a* 20   *b* 40; 3000
4 156   *5* 77; 19   *6* 35   *7* 74; no
8 b 50   *c* (*1*) no (*2*) yes (*3*) yes

Page 195   Exercise 4

1 a 1   *b* 0   *2 a* 1   *b* 0   *c* 1   *d* 0
3 a 1   *b* 0   *c* 0 and 1   *5 a* $\frac{1}{2}$ *b* $\frac{1}{2}$ *c* 1
6 a $\frac{1}{2}$ *b* $\frac{1}{2}$ *c* 1   *7 a* $\frac{5}{9}$ *b* $\frac{4}{9}$ *c* 1   *8* 1
9 $\frac{35}{36}$ *10* 0·65   *11* $\frac{2}{7}; \frac{5}{7}$

Page 197   Exercise 5B

1 a $\frac{1}{36}, \frac{1}{18}, \frac{1}{12}, \frac{1}{9}, \frac{5}{36}, \frac{1}{6}, \frac{5}{36}, \frac{1}{9}, \frac{1}{12}, \frac{1}{18}, \frac{1}{36}$ *b* no
2 7; 2 and 12   *3* $\frac{1}{18}$ *4* $\frac{1}{4}$ *5* $\frac{5}{12}$ *6* $\frac{5}{12}$
7 a $\frac{1}{6}$ *b* $\frac{1}{3}$ *8 a* $\frac{1}{3}$ *b* $\frac{4}{9}$ *c* $\frac{2}{3}$ *9* $\frac{5}{12}$

Page 199   Exercise 6B

1 $\frac{1}{36}$ *2* $\frac{1}{36}$ *3* $\frac{1}{12}$
4 a 12   *b* second row   *c* fourth column   *d* $\frac{1}{12}$ *e* $\frac{1}{2}, \frac{1}{6}$
5 e $\frac{1}{2}, \frac{1}{2}$ *f* $\frac{1}{4}$ *6* $\frac{1}{4}, \frac{1}{4}, \frac{1}{2}$ *7 a* $\frac{1}{4}$ *b* $\frac{1}{4}$ *c* $\frac{1}{4}$ *d* $\frac{1}{4}$; 1
8 a $\frac{1}{4}$ *b* $\frac{1}{2}$ *c* $\frac{1}{24}$ *d* 0   *e* $\frac{5}{12}$
9 Second row: THH, THT, TTH, TTT.
   a $\frac{1}{8}$ *b* $\frac{3}{8}$ *c* $\frac{1}{8}; \frac{1}{2}, \frac{1}{4}, \frac{1}{8}, \frac{1}{16}, \frac{1}{32}$ *10 a* $\frac{1}{5}$ *b* $\frac{8}{25}$

Answers

Arithmetic—Answers to Chapter 4

Page 204   Exercise 1
1. 3 h 10 min, 2 h 55 min
2. 9.45 am/5.10 pm; 11.50 pm/6.10 am; 10 am/5.20 pm; 11 pm/5.25 am
3. Dunkeld, 50 min; Pitlochry, 55 min
4. Perth, 30 min (both)
5. 7 h 25 min, 6 h 20 min, 7 h 20 min, 6 h 25 min
6. 6 h 25 min, 5 h 45 min, 6 h 10 min, 5 h 50 min
7. 0800, 1115, 1520, 1930, 2245, 0330
8. 0800, 2300, 1500 hour services
9. 10 h 42 min, 10 h 16 min, 11 h 44 min, 10 h 51 min, 11 h 24 min
10. 3 h 40 min, 3 h 20 min, 3 h 20 min, 3 h 20 min, 3 h 30 min

Page 207 Exercise 2
2. 3 h 20 min, 5 h, 27 km
3. 1340, 1215; 1140, 26 km from Glasgow
4. a 50 km, 50 km, 50 km/h   b 1036   c 0824
   d 60 km, 60 km/h   e 1036   f 0924, 70 km from Perth
5. about 700 km/h, 1110, 1500 km from Athens

Page 210   Exercise 3
1. a $37\frac{1}{2}$ km/h   b 8 m/s   c 70 km/h   d 12 cm/s
   e $12\frac{1}{2}$ m/min
2. a 192 km   b 4000 km   c 15 km   d $112\frac{1}{2}$ cm
   e 36 km
3. a 7 h   b 0·5 s   c 8 s   d 2 s   e 6 h 40 min
4. 115 km/h   5. 92 km/h   6. 378000 km
7. 50300 km   8. 8 h   9. 2·5 km   10. 45 km/h, 3 h
11. 5 min; 4·8, 14·4 km/h
12. a 2 h 57 min, 3 h 23 min, 3 h 11 min, 3 h 17 min   b 63, 53, 57, 56 km/h
13. 36 km/h   14. just over 8 min
15. 9×10^{12} km approximately   16. 0·13 s

Arithmetic—Answers to Revision Exercises

Page 214   Revision Exercise 1A
1. £12·84   2. £27·30   3. £29·40
4. a 21p, £2·10, £21   b $10\frac{1}{2}$p, £1·05, £10·50
   c $52\frac{1}{2}$p   d £1·19   e £1·$92\frac{1}{2}$
5. a £14·40   b £20·16   c £2·04   d £4·42

Answers

6 £9·63   7a £15   b £2·87   c £15
8 a £10650   b £124·20   c £93·08
9 a 8%   b $6\frac{2}{3}$%   c $6\frac{1}{4}$%
10a $16\frac{2}{3}$% profit   b $12\frac{1}{2}$% profit   c 20% loss
11a £96   b £64   c £40·80   d 0
12a 12%   b 125%   c 36%   d 130%
13a $16\frac{2}{3}$%   b 30%   c 5250   14 132000

Page 216   Revision Exercise 1B
1 £16·94   2 £14·25   3 £60, 0·67%   4 £32·23
5 Savings Bank by £2·10   6 50%   7 19%
8 a £360   b £129·60   c £156   d 0
9 £1·75   10 400%, 85%; 2·3%   11 5390, 4150

Page 217   Revision Exercise 2A
1 45p   2 80   3 40   4a £84   b 250
5 £6·60 6a 4·8 m   b 20·8 m   7 —   8 10 days
9 11   10a 1 : 20000   b 2·5 km   c 45 cm   11 144, £4
12a 100%, 71%, 44%, 25%   b 28, 70, 106, 127

Page 219   Revision Exercise 2B
1 £1·25, £90   2 27 min   3 £101·75   4 94·5 m
5 20 min   6 21 min   7 £648000   8a —   b 60 kg
9 a 750 kc/s   b 371 m
10 Table: 120 fr, 1200 fr; £27
    a 60, 1560, 2580   b £0·25, £10, £28·75
11a 75%, 45%, 69%   b 40, 58, 34
12a 7 : 9   b 3 : 1   c 3 : 19, 16%

Page 220   Revision Exercise 3A
1 a $\frac{1}{9}$ b $\frac{2}{9}$ c $\frac{1}{3}$ d $\frac{2}{3}$ 2a $\frac{13}{40}$ b $\frac{1}{5}$ c $\frac{19}{40}$
3 a $\frac{1}{5}$ b $\frac{4}{5}$ c $\frac{2}{5}$ 4a $\frac{1}{13}$ b 38 or 39
5 0·19, 0·46, 0·36   6 $\frac{1}{6}$, 60; $\frac{1}{4}$, 40
7 12 or 13, 18 or 19   8 430   9 $\frac{1}{9}$ 10 375, 250; $\frac{1}{2}$

Page 222   Revision Exercise 3B
1 a $\frac{3}{13}$ b $\frac{4}{13}$ c $\frac{7}{13}$ 2 $\frac{1}{20}$; 30   3 8
4 $\frac{1}{1000}$, 500   5a 3   b 0   c 18
6 a 12 or 13   b 37 or 38   7a $\frac{3}{16}$ b $\frac{1}{16}$ c $\frac{1}{4}$; 5, $\frac{1}{4}$
8 $\frac{1}{8}$ 9 $\frac{1}{15}$ 10a $\frac{1}{10}$ b $\frac{3}{5}$; $\frac{9}{25}$

Answers

Page 223   Revision Exercise 4

1   26 h, 20 h 20 min   *2*   1 h 10 min, 42 km/h   *3*   70 km/h
4 a   3 min   *b*   32 min   *5* $\frac{3}{4}$ h, $11\frac{1}{4}$ km   *6*   516 km/h
7   74 km/h, 18 12   *9*   2 h 34 min
10   13 08, 12 50, 12 31, 27 km from Stirling